HEYDEN INTERNATIONAL TOPICS IN SCIENCE

Editor: L. C. Thomas

RTL/2 Design and Philosophy	J. G. P. BARNES
Introduction to PASCAL	C. A. G. WEBSTER
Computer Applications in Ship Technology	C. KUO
An Introduction to Computer-based Library Systems	L. A. TEDD
Computers and Instrumentation—A Practical Handbook of Measurement, Interfacing and Control Circuits	A. CARRICK
Analysis with Ion-selective Electrodes	P. L. BAILEY
Analysis of Drugs of Abuse	E. BERMAN
Toxic Metals and their Analysis	E. BERMAN
Environmental Pollution Analysis	P. D. GOULDEN
Paper and Thin Layer Chromatographic Analysis of Environmental Toxicants	M. E. GETZ
Biometeorology—The Impact of the Weather and Climate on Humans and their Environment (Animals and Plants)	S. W. TROMP
Colour: its Measurement, Computation and Applications	G. J. CHAMBERLIN and D. G. CHAMBERLIN

PAPER AND THIN LAYER CHROMATOGRAPHIC ANALYSIS OF ENVIRONMENTAL TOXICANTS

PAPER AND THIN LAYER CHROMATOGRAPHIC ANALYSIS OF ENVIRONMENTAL TOXICANTS

M. E. GETZ

Agricultural Research Service
United States Department of
Agriculture, USA

HEYDEN

LONDON · PHILADELPHIA · RHEINE

Heyden & Son Ltd., Spectrum House, Hillview Gardens,
London NW4 2JQ, UK
Heyden & Son Inc., 247 South 41st Street, Philadelphia, PA 19104, USA
Heyden & Son GmbH, Münsterstrasse 22, 4440 Rheine, Germany

Getz, M E
Paper and thin layer chromatographic analysis of environmental toxicants.—(Heyden international topics in science).
1. Pollution—Measurement
2. Pollution—Toxicology
3. Environmental chemistry
4. Paper chromatography
5. Thin layer chromatography
I. Title
614.7 TD177 79-41719

ISBN 0-85501-451-2

Set by Eta Services (Typesetters) Ltd., Beccles, Suffolk
Printed litho and bound in Great Britain by W. & J. Mackay Ltd., Chatham

CONTENTS

FOREWORD

The object of this series of monographs is the timely dissemination of information on topics of current interest in science. Techniques, ideas and applications are described in sufficient detail to enable those who are not specialists in a particular subject to appreciate the applicability of the subject matter to their own work. Bibliographies included in these monographs will enable the reader to pursue the subject to any desired depth. The volumes will thus form compact definitive texts for the specialist also.

One topic of continuing current interest in science is that of the environment and its pollution. The present volume deals with the application of the relatively simple and inexpensive techniques of paper and thin layer chromatography to the detection, and determination, of toxic or potentially toxic contaminants of the environment, with special emphasis on the air we breathe, the food we eat, and the water we drink.

Because of the relatively cheap and simple methods involved, this volume will be of especial interest to analysts without access to more sophisticated and expensive equipment and, in particular, to those in developing countries. The extensive collection of chromatographic reference data (for animal drugs and feed additives, air and water pollutants, pesticides, herbicides, food additives, cosmetics and natural toxicants), which are included in this volume, make it a valuable laboratory reference source for all scientists working with these materials.

L. C. Thomas

PREFACE

'They shall rule the fish of the seas, the birds of the sky, the cattle, the whole earth and all the creeping things that creep on earth'.

Genesis: 1.26–27

Many centuries have passed since the above words were written. During this time man has not only used what nature has provided but has also, unfortunately, abused her. Today, in the twentieth century we are acutely aware of our environment. Pollution, ecology and environment are everyday words. There is much emphasis on the interaction between man and his surroundings, and the beneficial and detrimental effects of his influence on the environment.

As science and technology grow at an ever increasing rate, man has realized that though it is highly improbable that he could destroy the planet earth, he can certainly destroy those elements which allow him to dominate it. The introduction into nature of artificial chemical factors makes it necessary to ascertain if the environment is capable of diluting or altering these artifacts to prevent permanent detrimental effects. This need has heralded a new era in analytical chemistry. New separation techniques coupled with the availability of highly sophisticated electronic instrumentation have enabled the analyst to measure chemicals in ever smaller quantities.

Chromatographic techniques have facilitated the task of environmental analysis. This monograph is concerned with the liquid–liquid and liquid–solid adsorption phenomena as manifested in paper and thin layer chromatography. The two techniques are reviewed as analytical methods for monitoring natural and man-made pollutants and toxicants in the environment. The author, having selected the more significant topics from this vast subject, has tried to include enough pertinent information to enable an experienced analyst to develop a methodology for solving his particular problem.

Today, it is possible to make paper and thin layer chromatography as quantitative as other chromatographic methods by *in situ* determination of spot density. This may be accomplished with the help of optical scanning instruments by applying spectrophotometric methods of assay to the desired compounds after they have been resolved on a paper or thin layer chromatogram. It is no longer necessary to extract material from a migrated spot before quantifying it spectrophotometrically or otherwise.

Many of the methods described can be extended to be quantitative by the application of *in situ* optical scanning, which means that there are now three quantitative chromatographic techniques available to the analyst; gas–liquid chromatography with its many detectors, high performance liquid chromatography, and thin layer or paper chromatography with its range of chromogenic reagents.

I would like to thank my wife Eve for her constructive criticism and editing which always improves the content of my writings.

M. E. Getz

November 1979

CHAPTER 1

DRUGS AND ANIMAL FEED ADDITIVES

Man has been growing crops and raising animals for food for many thousands of years, but it has only been during the last two centuries that antibiotics and other drugs have been applied to the improvement of the many products of agronomy. One development of these applications has been the mixing of specific chemicals with animal feeds to increase resistance to disease and parasites in livestock.

Maintaining the health of the livestock population is important from two points of view. Firstly, it helps to ensure good quality meat, poultry, eggs and milk products for the consumer, at reasonable prices. Secondly, it greatly reduces the transmission of common diseases from animal to animal, and more importantly from animals to man.

Since the demand for animal protein has increased with the expanding world population, and agricultural practices are striving to keep production at a high level, antibiotics and other drugs have become increasingly important for maintaining efficient and competitive livestock industries. A number of drugs are used for the regulation of livestock growth and for the control of diseases such as coccidiosis, histomoniasis, worms and other epidemic diseases and parasites that can infect farm animals. Tables 1.1 and 1.2 list many of the chemicals that are used for these purposes in the United States.

The increasingly advanced technology of animal treatment has necessitated more complex regulation of these treatments by government. Any chemical used for the treatment of animals must now be shown to be both effective and safe for the animal, and to be free from harmful effects to the consumer. However, the possibility of drug residues in edible meat, milk, eggs, poultry and fish products always exists. This has meant that chemical analysis, which has always played an important role in governmental supervision of such chemical usage, has become increasingly necessary as the chemicals used in animal husbandry continually change.

TABLE 1.1
Antibiotics commonly used for animal treatment

Bacitracin	Neomycin
Chlortetracycline	Novobiocin
Erythromycin	Nystatin
Griseofulvin	Oleandomycin
Hygromycin B	Oxytetracycline
Lincomycin	Penicillin G
Monensin	Tylosin

Paper and thin layer chromatography can be useful tools for determining the identity and amount of chemicals in animal feed, and for monitoring the amounts of chemical residues in edible tissues and products.

1.1 EXTRACTION AND ISOLATION

The first step in any trace analytical procedure is the extraction of the desired chemical from the matrix in which it is contained. This can be done by various means such as:

(1) Stripping, used in the extraction of Novobiocin from feed.[1] A 1 g sample is placed in a stoppered graduated cylinder or flask, and extracted twice by vigorous shaking with ethyl acetate. The supernatant liquid is decanted and diluted to volume for microbiological assay or further purified for chromatographic separation.
(2) Soxhlet extraction, applied to extracting Griseofulvin from a feed sample.[2] A few grams of the finely-ground sample are placed in a fat-free thimble and extracted with chloroform for 2 h. The chloroform extract is evaporated down to a small volume and then diluted to 100 ml for further analytical treatment.
(3) High speed homogenization, whereby Clopidol residues are extracted from muscle, liver and kidney[3] with methanol and HYflo Super-Cel. The homogenate is filtered and the methanol filtrate is saved for further isolation steps.

After the relevant compound or compounds have been extracted, it is usually necessary to render them pure enough to be chromatographed with a minimum of interference from other co-extractives. This step is commonly referred to as the clean-up procedure, and experience has shown that a very efficient clean-up is usually necessary to prepare a sample for analysis by paper or thin layer chromatography. A widely applied bipartite clean-up method employs liquid–liquid partitioning between polar and nonpolar solvents,

TABLE 1.2
Drugs commonly used on animals in the United States

Common name[a]	Chemical name
Aklomide	2-Chloro-4-nitrobenzamide
Amprolium	1-(4-Amino-2-*n*-propyl-5-pyrimidinyl)-2-picolinium chloride
Arsanilic acid	*p*-Aminobenzenarsonic acid
Buquinolate	Ethyl-4-hydroxy-6,7-diisobutoxy-3-quinolinecarboxylate
Decoquinate	Ethyl-6-*n*-decyloxy-7-ethoxy-4-hydroxy-3-quinolinecarboxylate
DES	Diethylstilbesterol
Dimetridazole	1,2-Dimethyl-5-nitroimidazole
Enheptin	2-Amino-5-nitrothiazole
Ethopabate	Methyl-4-acetamido-2-ethoxybenzoate
Furazolidone	*N*-(5-Nitro-2-furfurylidene)-3-amino-2-oxazolidone
Glycarbylamide	4,5-Imidazoledicarboxamide
Ipronidazole	1-Methyl-2-isopropyl-5-nitroimidazole
Nequinate	Methyl-7-benzyloxy-6-butyl-1,4-dihydro-4-oxo-3-quinolinecarboxylate
Nicarbazin	4,4′-Dinitrocarbanilide-2-hydroxy-4,6-dimethylpyrimidine
Nifursol	3,5-Dinitrosalicylic acid
Nihydrazone	5-Nitro-2-furfuraldehyde acetylhydrazone
Nithiazide	1-Ethyl-3-(5-nitro-2-thiazolyl)urea
Nitrodan	3-Methyl-5-(*p*-nitrophenyl-azo-rhodanine)
Nitrofurazone	5-Nitro-2-furaldehyde semicarbazone
Nitromide	3,5-Dinitrobenzamide
Nitrophenide	*m,m*′-Dinitrodiphenyldisulfide
Nitarsone	4-Nitrophenylarsonic acid
Phenothiazine	
Piperazine	
Racephenicol	DL-Threo-2,2-dichloro-*N*-(β-hydroxy-α-hydroxymethyl)-*p*-(methylsulfonyl)phenethyl acetamide
Reserpine	
Roxarsone	4-Hydroxy-3-nitrobenzenearsonic acid
Sulfathiazole	
Sulfamerazine	
Sulfadimethoxine	2,5-Diamino-5-(4,5-dimethoxy)methylbenzyl pyrimidine
Sulfaguanidine	
Sulfaquinoxaline	
Sulfamethazine	*N*-(4,6-Dimethyl-2-pyrimidinyl)sulfanilamide
Sulfanitran	Acetyl-(*p*-nitrophenyl)sulfanilamide
Thiobendazole	2-(4-Thiazolyl)benzimidazole
Zoalene	3,5-Dinitro-*o*-toluamide

[a] Where the common and chemical names are identical the chemical name is not repeated.

followed by column chromatography on adsorbents such as alumina, silica gel or Florisil.

There are two major steps in preparing a sample for analysis. The compound must first be extracted from the containing matrix and then separated from unwanted co-extractives so that there is little or no interference when developing the thin layer or paper chromatogram.

1.2 ANALYSIS OF ANTIBIOTICS

Antibiotics are used for promoting the growth of livestock and poultry, and it is generally accepted that they arc capable of three major types of biological action.[4]

(1) A metabolic effect, in which the antibiotic affects the metabolism of the animal favorably, so that it grows at a faster rate.
(2) The suppression of undesirable organisms, which compete with the animal for nutrients, and the stimulation of desirable organisms.
(3) The suppression of organisms that cause disease.

1.2.1 Bacitracin

The antibiotic bacitracin is usd as a growth promoter, as an inhibitor of liver abscesses, for increasing egg production and hatchability, for suppressing enteritis and chronic respiratory diseases, and for minimizing the effects of stress.

A successful thin layer separation (TLC) has been obtained by Ikekawa *et al.*[5]

Adsorbent: Silica Gel G
Developing solvents: (1) ethanol–ammonium hydroxide–water (8:1:1)
(2) ethanol–water (4:1)

R_F: (1) =0.58; (2) =0.13.

The compound can be made visible with ninhydrin, or UV light or by bio-autography.

Nussbaumer[6] also resolved bacitracins A and B by TLC.

Adsorbent: Silica Gel G
Developing solvent: *n*-butanol–acetic acid–water (4:1:2)

Bacitracin A, $R_F = 0.35$; bacitracin B, $R_F = 0.22$.

The spots were detected in UV light after heating the chromatogram to 110 °C, or with ninhydrin dissolved in a butanol–acetic acid solution.

1.2.2 Chlortetracycline and Oxytetracycline

Chlortetracycline is used as a growth promoter, as a preventive against anaplasmosis, bacterial pneumonia, foot rot, liver abscesses, respiratory infections, shipping fever, enteritis, coccidiosis, synovitis, atrophic rhinitis and cervical abscesses, and for increasing egg production and hatchability. Oxytetracycline is used to increase milk production, to produce better eggshell texture, and to prevent blue comb, cholera, hepatitis, bacterial diarrhea, enterotoxemia, leptospirosis, and hexamitiasis in animals and man.

Paper chromatography can be used to resolve these two antibiotics. One approach has been to develop the paper chromatogram with a solvent mixture of *n*-butanol, acetic acid and water (4:1:5).[7] The migrated spots were detected by bioautography and the R_F values for oxytetracycline and chlortetracycline were 0.60 and 0.69, respectively.

Another approach was to impregnate the paper with pH 3.5 McIlvaine's citrate–phosphate buffer,[8] and then to develop the chromatogram with a solvent system consisting of chloroform, nitromethane and pyridine (10:20:30), using bioautography to detect the spots. (R_F oxytetracycline = 0.13; R_F chlortetracycline = 0.50.)

Kelly and Buyska[9] impregnated the chromatographic paper with 0.1M EDTA solution and then developed with a mixture of *n*-butanol, acetic acid and water (4:1:5). This mixture forms two phases of which the upper phase is used for development. Bioautography is used for detection. This solvent system gave an R_F of 0.59 for oxytetracycline and an R_F of 0.76 for chlortetracycline. When the upper phase of the solvent mixture, *n*-butanol–ammonium hydroxide–water (4:1:5) was used to develop the chromatogram oxytetracycline and chlortetracycline had R_F values of 0.27 and 0.47, respectively.

Two other useful paper chromatographic systems were developed by Bohonas *et al.*,[10] and Hickey and Phillips.[11] The former soaked chromatographic paper with 0.3M pH 3 phosphate buffer, air-dried the paper and then developed in *n*-butanol saturated with water. Bioautography was used for detection; oxytetracycline gave an R_F of 0.40 and chlortetracycline an R_F of 0.65. The latter group chromatographed chlortetracycline without using a buffer but included acetic acid in the developing solvent.

Adsorbent: Schleicher and Schuell No. 507 chromatographic paper

Developing solvent: acetic acid–*n*-butanol–water (1:2:1)

R_F = 0.77.

Detection technique: bioautography

When TLC was applied to the resolution of the two tetracyclines some difficulties were apparent in the separations. However, Gyanchandani *et al.*[12] were able to develop a system for separating them.

The thin layer was prepared by mixing 50 g of celite with a 0.1M aqueous solution of EDTA plus a 20% polyethylene glycol 400 solution in glycerine (95:5). The plates were coated with a layer of 0.25 mm thickness, and dried overnight at room temperature. The following two developing solutions gave good resolution.

(1) Methyl ethyl ketone saturated with pH 4.7 McIlvaine's buffer.

R_F oxytetracycline = 0.60; R_F chlortetracycline = 0.76.

(2) Dichloromethane–ethyl formate–ethanol (9:9:2) saturated with pH 4.7 McIlvaine's buffer.

R_F oxytetracycline = 0.20; R_F chlortetracycline = 0.60.

Two chromogenic reagents gave good results: Fast Blue B (a diazonium salt), under alkaline conditions, gave a pink spot with chlortetracycline and a yellow spot with oxytetracycline; a 50 mg per cent solution of diphenylpicrylhydrazyl sprayed under alkaline conditions gave a yellowish pink spot for chlortetracycline and a brownish yellow spot for oxytetracycline.

1.2.3 Erythromycin

This is used to promote growth, and to prevent respiratory diseases, coryza and stress. Majumdar and Majumdar[13] developed a paper chromatographic procedure for the resolution of erythromycin isomers, where the paper was developed with methanol, water and acetone (19:75:6), and the resulting spots were detected with bioautography using *Micrococcus pyogenes var. Aureus*, or *Bacillus subtilis* (erythromycin A, $R_F = 0.7$, and erythromycin B, $R_F = 0.6$).

A thin layer separation was accomplished[14] by coating a plate with a slurry of Silica Gel G in a 0.02N aqueous sodium acetate solution, developing the plate in methanol and 0.02N sodium acetate (120:30), then detecting the migrated compound with a solution of glucose, 85% phosphoric acid, water, ethanol and *n*-butanol (2 g + 10 ml + 40 ml + 30 ml + 30 ml). After spraying with the reagent solution the plate was heated at 150 °C for 5 min, erythromycin formed a blue grey color. ($R_F = 0.28$.)

1.2.4 Griseofulvin

This is a fungicide which is used for the systemic treatment of athletes foot in humans and also serves as a general fungicidal treatment for animals. It can be chromatographed using paper in conjunction with a number of solvent systems.[15] The paper may be:

(i) impregnated with chloroform and developed with a butanol-saturated water solution. ($R_F = 0.22$.)
(ii) impregnated with ethyl acetate and developed with butanol-saturated water. ($R_F = 0.22$.)
(iii) impregnated with ethyl acetate and developed with a butanol, ethanol and water mixture (5:1:4). ($R_F = 0.37$.)
(iv) impregnated with methyl ethyl ketone and developed with butanol-saturated water. (R_F value $= 0.24$.)

Detection of the migrated compound was carried out by UV irradiation. Thin layer techniques can also be used for the chromatography of Griseofulvin.[16] A plate is coated with a layer of Silica Gel HF254 and developed in two different solvent systems.

(1) Butyl acetate–acetone (4:1, $R_F = 0.49$)
(2) Chloroform–methanol (98:2, $R_F = 0.68$)

The migrated spot may be detected with a solution of 5% potassium carbonate and 0.2% potassium permanganate.

1.2.5 Lincomycin

This is used primarily for promoting growth in animals, and can also be separated by paper chromatography:[17] the chromatogram is developed with a solution of 1-butanol, acetic acid and water (2:1:1), and the antibiotic then detected by bioautography using *Sarcina lutea*. ($R_F = 0.70$).

Silica Gel G was used as adsorbent for the thin layer chromatography of lincomycin.[18] The chromatogram was developed in ethyl acetate, acetone and water (8:5:1), and the spot detected by bioautography using *Sarcina lutea*. ($R_F = 0.3$.)

1.2.6 Monensin

This antibiotic is used for the control of coccidiosis. The various types of monensin have been resolved by paper chromatography,[19] using Whatman No. 1 paper developed in a water, methanol, acetic acid benzene mixture (72:24:0.5:3.5). The migrated spots were detected by bioautography using *Bacillus subtilis ATCC 6633*. (Monensin A, $R_F = 0.50$; Monensin B, $R_F = 0.90$; Monensin C, $R_F = 0.35$.)

Thin layer chromatography of monensin was carried out with silica gel as adsorbent; the developing solvent was ethyl acetate, and the migrated compounds were detected by spraying with a 3% vanillin solution in 1.5% ethanolic sulfuric acid and heating at 100 °C until spots were a bright red color. (Monensin A, $R_F = 0.5$; Monensin B, $R_F = 0.5$; Monensin C, $R_F = 0.2$.)

1.2.7 Neomycin

This is used to treat bacterial enteritis, enterotoxaemia and dysentery. Paper chromatographic separation was obtained[19] by developing the chromatogram in a solvent mixture of *n*-propanol, glacial acetic acid and water (9:1:10). Neomycin B and Neomycin C gave R_F values of 0.26 and 0.30, respectively. They were both detected with ninhydrin. Another paper method[20] used methyl ethyl ketone, *t*-butanol, methanol and 6.5N ammonium hydroxide (16:3:1:6) as the developing solvent, and ninhydrin as the detecting agent. (Neomycin B, $R_F = 0.54$; Neomycin C, $R_F = 0.30$.) The R_F values were calculated relative to the R_F value for neamine, which was taken as 1.00.

An interesting approach to the thin layer separation of Neomycin was devised by Brodasky.[21] Plates were prepared from Nuchar activated carbon by making a slurry of the following composition: Nuchar C-190N + calcium sulfate ($CaSO4 . \frac{1}{2}H_2O$) + water (30 g + 1.5 g + 220 ml). This was applied to glass plates immediately after mixing and allowed to air-dry. Coated plates were stored for 16 h before use. Chromatograms were then developed with a solution of 0.5N sulfuric acid, neutralized by exposure to ammonia fumes for

5 min, and the antibiotic was detected with bioautography using the organism *Bacillus pumilis*. The antibiotics appeared as white spots against the black background of the charcoal. (Neomycin A, $R_F = 0.61$; Neomycin B, $R_F = 0.21$; Neomycin C, $R_F = 0.43$.)

1.2.8 Novobiocin

This is used to prevent blisters on the udders, and synovitis and fowl cholera. Paper chromatographic separation can be achieved by impregnating the paper with an immobile phase of capryl alcohol and methanol (1:5), and developing with a mobile phase of 0.1M phosphate buffer equilibrated with capryl alcohol.[22] Bioautography is used for detection ($R_F = 0.25$).

Using a Silica Gel G layer as adsorbent in TLC of novobiocin, a developing solvent mixture of isoamyl acetate, methanol, formic acid and water (65:20:5:10), and detection by bioautography an R_F value of 0.59 was obtained.[23]

1.2.9 Nystatin

Used for the prevention of mycosis and mycotic diarrhea, nystatin can be separated by paper chromatography,[23] using Whatman No. 1 paper developed with butanol-saturated water as solvent. The compound is then detected by bioautography with *Saccharomyces carlsbergensis K-20* or with a solution of glucose and 2,3,5-triphenyltetrazolium chloride. ($R_F = 0.22$.) Silica Gel G[24] can be used for thin layer chromatography, developing the chromatogram in *n*-butanol, acetic acid and water (4:1:2), and detecting nystatin with a 0.02N solution of potassium permanganate. ($R_F = 0.45$.)

1.2.10 Oleandomycin

This is used as growth promoter. It may be chromatographed on paper[25] by using an immobile phase of formamide and developing with two mobile phases:

(1) benzene–chloroform (3:1). ($R_F = 0.32$.)
(2) benzene–chloroform (1:1). ($R_F = 0.63$.)

The antibiotic can be detected by bioautography using *Bacillus subtilis*.

Silica gel adsorbent was used for TLC,[26] the plate being developed in a butanol, acetic acid and water mixture (3:1:1). The migrated spot was made visible by charring with sulfuric acid or by bioautography with *Bacillus subtilis*. ($R_F = 0.53$.)

1.2.11 Penicillin G

This is the form of penicillin used for promoting growth, and for the prevention of bloat and blue comb. In order to detect it by paper chromatography the paper was impregnated with a 2% liquid paraffin solution in ethyl ether and air-dried.[27] The mobile phase was butyl acetate equilibrated with the paraffin solution. The compound was detected by bioautography with *Bacillus subtilis* $R_F = 0.48$.

1.2.12 Tylosin

This is used to promote growth and in the treatment of chronic respiratory diseases. It can be separated on chromatographic paper[28] impregnated with a pH 4 buffer solution and developed in methyl ethyl ketone. ($R_F = 0.46$.) The R_F value using paper not impregnated with buffer was 0.81. Another developing agent used was an aqueous solution containing 7% sodium chloride and 2.5% methyl ethyl ketone, ($R_F = 0.57$).

Tylosin has also been separated using thin layer chromatography.[29]

Adsorbent: Silica Gel G
Developing solvents: chloroform–acetone (60:40). ($R_F = 0.12$)
ethyl acetate–methanol (85:15). ($R_F = 0.60$)
Detection: bioautography.

1.3 ANALYSIS OF SULFONAMIDES

Sulfonamides are chemotherapeutic agents for the treatment of bacterial infections. They are substituted amides of sulfanilic acid which are selectively toxic; their growth-inhibiting effect on bacterial organisms allows the body's defenses to overcome an infection.

Figure 1.1 shows the structures of four sulfonamides: sulfaquinoxaline, sulfamerazine, sulfamethazine and sulfaguanidine. Sulfaquinoxaline is used for the treatment of coccidiosis in chickens and turkeys, cholera typhoid in fowl, bacterial enteritis in swine, and shipping dysentery and coccidiosis in lambs. Sulfamerazine is used for the treatment and control of acute mastitis, shipping fever, and pneumonia in cattle, equine influenza, canine respiratory infection, infectious pododermatitis in cattle, urinary infections and fowl typhoid. The treatment of acute mastitis, shipping fever, pneumonia in cattle,

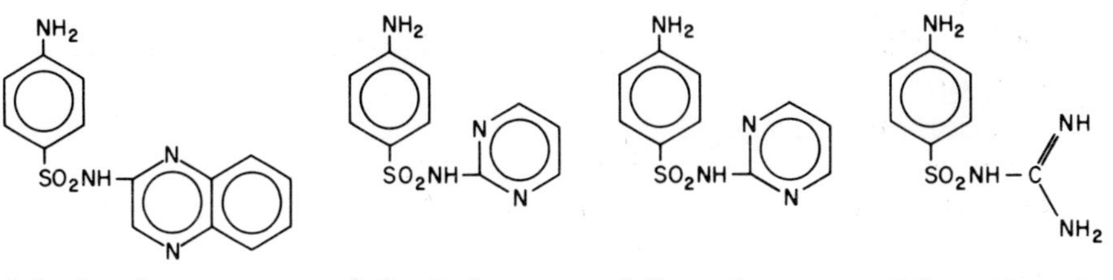

Fig. 1.1. Structural formulae of some typical sulfonamides.

equine influenza and canine respiratory infections can also be achieved with sulfamethazine, while sulfaguanidine is restricted to the treatment of mammalian coccidiosis and swine enteritis.

1.3.1 Separation of sulfonamides

Paper chromatography[30] using different developing solvent systems has been applied to the separation of some sulfonamides.

Solvent system (1): isopropanol–25% ammonia (4:1)

Sulfaguanidine, $R_F = 0.48$; sulfathiazole, $R_F = 0.42$; sulfamerazine, $R_F = 0.31$.

Solvent system (2): chloroform–methanol–25% ammonia (80:45:15)

Sulfaguanidine, $R_F = 0.54$; sulfathiazole, $R_F = 0.48$; sulfamerazine, $R_F = 0.52$.

Solvent system (3): isoamyl acetate–pyridine–25% ammonia (14:15:4)

Sulfaguanidine, $R_F = 0.46$; sulfathiazole, $R_F = 0.25$; sulfamerazine, $R_F = 0.26$.

Solvent system (4): chloroform–acetone–25% ammonia (18:100:20)

Sulfaguanidine, $R_F = 0.76$; sulfathiazole, $R_F = 0.50$; sulfamerazine, $R_F = 0.27$.

Detecting reagents: fluorescent indicators sprayed onto the paper renders the sulfonamides visible as quenched areas, or the paper may be sprayed with sodium nitrite, in hydrochloric acid solution, to form diazonium compounds and then reacted with *N*-(1-naphthyl)ethylenediamine.

The same investigators also used thin layer chromatographic techniques with silica gel and alumina as the adsorbents.

Adsorbent: Silica Gel G

Solvent (1) ethyl acetate–methanol–25% ammonia (85:15:15)

Sulfaguanidine, $R_F = 0.50$; sulfathiazole, $R_F = 0.80$; sulfamerazine, $R_F = 0.59$.

Solvent (2) methyl isobutyl ketone–acetone–25% ammonia (85:15:15)

Sulfaguanidine, $R_F = 0.64$; sulfathiazole, $R_F = 0.57$; sulfamerazine, $R_F = 0.45$.

Adsorbent: Alumina G

Solvent: ethyl acetate–methanol–25% ammonia (85:15:15)

Sulfaguanidine, $R_F = 0.78$; sulfathiazole, $R_F = 0.42$; sulfamerazine, $R_F = 0.25$.

Robinson[31] separated the sulfonamides by paper chromatography, developing in a solvent system consisting of *n*-butanol–ammonium hydroxide–water

(4:1:3). (Sulfaguanidine, $R_F = 0.36$, sulfamerazine, $R_F = 0.50$ and sulfamethazine $R_F = 0.48$.)

Another solvent system consisted of water-saturated *n*-butanol and pyridine (8:1). (Sulfaguanidine, $R_F = 0.30$, sulfamerazine, $R_F = 0.44$ and sulfamethazine, $R_F = 0.65$.) The spots were made visible with Ehrlich's reagent.[32]

Karpitschka[33] used polyamide layers to separate the sulfonamides. The developing solvent consisted of a mixture of petroleum ether, chloroform and butanol (1:1:1). (Sulfaguanidine, $R_F = 0.15$, sulfamerazine, $R_F = 0.44$ and sulfamethazine, $R_F = 0.52$.)

1.4 ANALYSIS OF NITROFURANS

Nitrofurans are synthetic compounds which have an antimicrobial action. They have in common the 5-nitro group, which has to be present if they are to be effective against microorganisms. They have a broad range of activity and are only slightly toxic to humans. It appears that their main effect is through enzyme inhibition in the microorganism. Figure 1.2 shows structures of some commonly used nitrofurans.

1.4.1 Nitrofurazone

This is used for the treatment of wounds, and the control of mastitis, necrotic enteritis in hogs, genital tract infections in cattle and horses, coccidiosis,

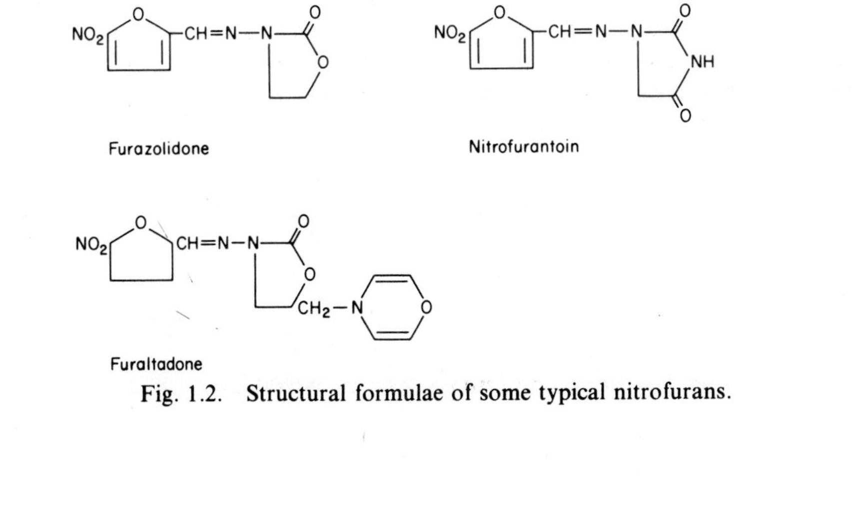

Fig. 1.2. Structural formulae of some typical nitrofurans.

typhoid and pullorum in chickens. Antkowiak and Spatorico[34] chromatographed a series of nitrofuran compounds with Alumina G. Nitrofurazone may be resolved by development with acetone and acetic acid (100:2) and made visible by treatment with a base and viewing the fluorescence at 254 nm. The plate could also be treated with 4-methylumbelliferone and ammonia vapor and the spot observed at 254 nm. ($R_F = 0.43$.)

1.4.2 Furazolidone

This is used for treating necrotic enteritis and vibrionic dysentery in hogs, and coccidiosis, typhoid and pullorum in chickens. Breinlich[35] was able to chromatograph furazolidone with paper chromatography and a developing solution of *n*-butanol, acetic acid and water (4:1:5). The R_F value was found to be 0.67 after treatment with a base and observing the fluorescent spot under UV light. He also applied thin layer chromatography, using a Silica Gel G plate, developing with chloroform and diethylamine (9:1) and detecting the spot by ultraviolet irradiation. ($R_F = 0.45$.)

Alumina G was also used by Antkowiak and Spatorico[34] to resolve furazolidone. ($R_F = 0.84$.)

1.4.3 Nihydrazone

This is used for the prevention and treatment of coccidiosis, typhoid, pullorum and CRD aerosaccullitis in chickens. It can be chromatographed on Alumina G[34] and gives an R_F value of 0.64.

1.5 ANALYSIS OF ARSENICALS

Arsenic has been used in human and animal medication in the past. It is present in plants and animal tissues at various concentrations; milk has an average concentration of 0.01 ppm, while fish, shellfish and crustaceans have the highest averages of 1.4–174 ppm. It has been found that organoarsenicals are effective growth promoters of pigs, chickens and turkeys. When properly used no significant increase is found in the intake of arsenic by man. Figure 1.3 shows some of the arsenicals in current use.

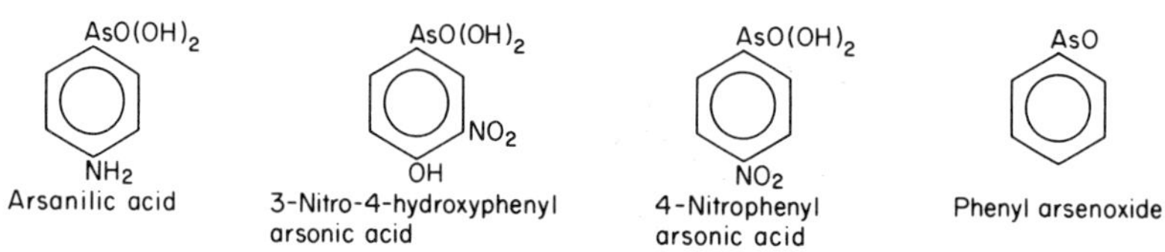

Fig. 1.3. Structural formulae of some typical arsenicals.

1.5.1 Resolution of arsenicals

Polyamide thin layer chromatography[34] with a solvent system of *n*-butanol, water and acetic acid (80:20:2.5) has been useful in their resolution. They are made visible by observation under UV light with a fluorescent indicator.

(Arsanilic acid, $R_F = 0.69$; carbarsone, $R_F = 0.54$; nitarsone, $R_F = 0.46$; 3-nitro-4-hydroxyphenylarsonic acid, $R_F = 0.36$.)

1.6 ANALYSIS OF OTHER DRUGS

The use of coccidiostats, histomonostats and antihelminthics in poultry feeds has helped to prevent the loss of poultry to these diseases and the resulting increase in flocks has made poultry one of the cheapest meats in the United States.

Some of these compounds have been chromatographed with thin layer chromatography.[34]

1.6.1 Differentiation of nicarbazine, nitrophenide and zoalene

Adsorbent: Silica Gel G

Solvent: ethyl ether–dimethyl sulfoxide–*n*-butanol–acetic acid (98:1:1:2).

Nicarbazine, $R_F = 0.25$; zoalene, $R_F = 0.73$; nitrophenide, $R_F = 0.95$.

The R_F of nitrophenide could be brought down to 0.80 if ethyl acetate and acetic acid (75:25) was used as the developing solvent.

1.6.2 Ethopabate

This gave an R_F value of 0.25 when hexane, acetone and *n*-butanol (21:2:2) was used as the developing solvent with Silica Gel G.

1.6.3 Differentiation of amprolium, phenothiazine, piperazine and sulfaquinoxaline

Adsorbent: Aluminium Oxide G

Developing solvent: ethanol–ammonium hydroxide (80:20)

Phenothiazine, $R_F = 0.81$; amprolium, $R_F = 0.16$; piperazine, $R_F = 0.58$; and sulfaquinoxaline, $R_F = 0.68$.

Amprolium can also be chromatographed on polyamide layers[35] and a solvent system consisting of ethyl ether, ethanol and acetic acid (96:3:1). The R_F value given by this method was 0.35.

It can now be seen that paper and thin layer chromatography can be of use for monitoring drugs used in treatment of animals reared for consumption.

REFERENCES

1. *Official Methods of Analysis*, 12th Edn, sect. 42.236, AOAC, Washington DC, 1975.
2. *Official Methods of Analysis*, 12th Edn, sect. 42.275, AOAC, Washington DC, 1975.
3. *Official Methods of Analysis*, 12th Edn. sect. 41.018, AOAC, Washington DC, 1975.
4. *The Use of Drugs in Animal Feeds*, National Academy of Sciences, Washington DC, 1969.
5. T. Ikekawa, F. Iwami, E. Akita and H. Umezawa, *Antibiotics (Tokyo) Ser. A* **16**, 56 (1963).
6. P. A. Nussbaumer, *Pharm. Acta Helv.* **40**, 210 (1965).
7. P. P. Regna and I. A. Solomons, *Ann. N.Y. Acad. Sci.* **53**, 229 (1950).
8. G. B. Selzer and W. W. Wright, *Antibiot. Chemother.* **7**, 292 (1957).
9. R. G. Kelly and D. A. Buyska, *Antibiot. Chemother.* **10**, 604 (1960).
10. N. Bohonas, A. C. Dornbush, L. I. Feldman, J. H. Martin, E. Pelcak and J. H. Williams, *Antibiot. Ann.* 49–55 (1953–1954).
11. R. J. Hickey and W. F. Philips, *Anal. Chem.* **26**, 1640 (1954).
12. N. D. Gyanchandani, I. J. McGilvery and D. W. Hughes, *J. Pharm. Sci.* **59**, 224 (1970).
13. M. K. Majumdar and S. K. Majumdar, *Appl. Microbiol.* **17**, 763 (1969).
14. C. W. Pettinga, W. M. Starke and F. R. van Abeele, *J. Am. Chem. Soc.* **76**, 569 (1954).
15. E. G. McNall, *Arch. Dermatol.* **81**, 657 (1960).
16. A. Segal and E. H. Taylor, *J. Pharm. Sci.* **57**, 874 (1968).
17. D. J. Mason, A. Dietz and C. Deboer, *Antimicrob. Agents Chemother.* 554–559 (1962–1963).
18. A. D. Argoudelis, J. H. Coats and B. J. Magerlein, *J. Antibiot.* **25**, 191 (1972).
19. M. E. Haney Jr and M. M. Moehn, *Antimicrob. Agents Chemother.* 349 (1967–1968).
20. K. L. Rinehart Jr, A. D. Argoudelis, W. A. Goss, A. Sohler and C. P. Schaffner, *J. Am. Chem. Soc* **82**, 3938 (1960).
21. T. F. Brodasky, *Anal. Chem.* **35**, 342 (1963).
22. E. J. Wolf and R. Nescot, *Antibiot. Ann.* 1035–1039 (1956–1957).
23. J. Burns and D. F. Holtman, *Antibiot. Chemother.* **9**, 398 (1959).
24. P. A. Mussbaumer, *Pharm. Acta Helv.* **43**, 462 (1968).
25. K. Murai, B. A. Sabin, W. B. Celmer and F. W. Tanner, *Antibiot. Chemother.* **9**, 485 (1959).
26. M. V. Kalinina and E. I. Surokova, *Antibiotika* **13**, 112 (1968).
27. T. Watanabe, S. Endo and Y. Iida, *Antibiotics* **15**, 112 (1962).
28. R. L. Hamill and W. M. Stark, *Antibiotics* **17**, 133 (1964).
29. M. Debockee and K. J. Baeten, *J. Chromatogr.* **61**, 125 (1971).
30. M. T. Van der Venne and J. B. T'Siobel, *J. Pharm. Belg.* **19**, 577 (1963).
31. R. Robinson, *Nature*, **168**, 512 (1951).
32. P. L. de Reeder, *Anal. Chim. Acta* **8**, 325 (1953).
33. N. Karpitschka, *Mikrochim. Acta* 157 (1963).
34. J. J. Antkowiak and A. L. Spatorico, *J. Chromatogr.* **29**, 277 (1967).
35. J. Breinlich, *Dtsch. Apothek. Ztg* **104**, 535 (1964).

CHAPTER 2

AIR AND WATER POLLUTION

Water and air are both vital for the support of all forms of terrestrial life. Water is involved in a cyclical process between the vapor state and the liquid or solid state. There appears to be a constant amount available in the biosphere, but the vapor in the atmosphere shifts to different areas of the planet due to gravitational and temperature changes, and it is this shifting which causes weather changes. The gases oxygen, carbon dioxide and nitrogen are similarly involved in cyclical processes, of which some are vital to the maintenance, breakdown and recycling of life forms.

During this century, the industrialized nations of the world have gradually recognized the detrimental effects that can result from the indiscriminate disposal of waste products into bodies of water and into the atmosphere. It is hoped that the newly industrialized nations of the world will learn from the experiences of the developed nations. They need to be concerned not only with the raising of living standards for their people, but also with doing so in a manner which will have a minimal effect on the environment.

International agreements develop slowly, but eventually a constant monitoring of the earth's water sources and atmosphere will be necessary to ascertain the types and amounts of waste products being released into them. This information may then show the need for an international committee with the powers necessary to rectify any dangerous situations that develop in the environment. Analytical chemistry has played, and will continue to play, a significant role in such monitoring of the environment, and paper and thin layer chromatography are two simple techniques that contribute towards identifying and quantifying many of the materials in the air and in water sources.

2.1 CARCINOGENS IN THE AIR

Air pollution can be caused by combustion from industrial manufacturing, automobile emissions, heating and cooking sources and natural phenomena. Particulate matter is formed which can be either solid or fine liquid droplets, which becomes airborne so that it may be displaced and carried to sites far removed from whence it originated. In the case of natural air pollution, volatile chemicals may be emitted by vegetation, forest fires and volcanic eruptions.

Table 2.1 lists some of the organocyclic compounds that have carcinogenic effects in man and animals.

TABLE 2.1
Carcinogenic polycyclic hydrocarbons

Benz[*a*]anthracene	2-Aminofluorene
1,2,5,6-Dibenzanthracene	1,2,5,6-Dibenzofluorene
6-Methyl-1,2-benzanthracene	1,2,3,4-Dibenzophenanthrene
10-Methyl-1,2-benzanthracene	3-Methyl cholanthrene
5,9,10-Trimethyl-1,2-benzanthracene	5,6-Dimethyl chrysene
6,9,10-Trimethyl-1,2-benzanthracene	1,2,5,6-Dibenzcarbazole
Benzo[*a*]pyrene	3,4,5,6-Dibenzcarbazole
3,4,8,9-Dibenzpyrene	1,2,5,6-Dibenzacridine
Benzo[*b*]fluoranthene	1,2,7,8-Dibenzacridine
Benzo[*i,j*]fluoranthene	Indeno[1,2,3-*cd*]pyrene

2.1.1 Extraction methods

Many attempts have been made to extract organic compounds efficiently from air samples. They are usually based on two general techniques, one of which is to sample air continuously, by bubbling it through a solvent for a specific period of time, and to measure the volume of air with a suitable flow meter. The amount of polycyclic hydrocarbons found can then be expressed in concentration per volume and unit time. The other method is to allow the air-flow to impinge upon or pass through a solid adsorbent such as activated charcoal or a fiberglass filter. The adsorbed polynuclear compounds are then displaced or extracted with a suitable solvent. Solvents that have given good extraction recoveries are cyclohexane, benzene, chloroform, acetone and methanol.

There is still a great deal of research being done on finding the best approach to the collection and extraction of airborne pollutants. Attempts have been made to standardize methods, as shown in the publications of the United States Intersociety Committees.[1,2] The problems encountered in sampling airborne pollutants, and methods devised to overcome these problems, are described in a recent book by Goulden.[3]

The following steps could serve as guidelines for analysts wishing to develop a procedure for specific problems. Standard mixtures of the carcinogens of interest could be impregnated onto a carrier such as celite and blown through an adsorbent or solvent to simulate particulate matter. This should be done in an isolated room or container so that there is no danger to the experimenter.

When a solvent is used the volume may be reduced to concentrate the solution for paper or thin layer chromatography. If a solid adsorbent or filter is used the carcinogen may be extracted with a range of solvents to find the most suitable one for the particular problem. These tests would determine the percentage recovery of the polycyclic compounds under investigation. Application of these methods to actual field samples may necessitate a further isolation step to eliminate interferences, which could be accomplished by other chromatographic techniques such as liquid/liquid or liquid/solid partitioning.

2.1.2 Paper chromatography

Dubois *et al.*[4] applied descending chromatography to the separation of some polycyclic hydrocarbons. They used Schleicher and Schuell No. 2043b acetylated paper, with a developing solvent of toluene, methanol and water (1:10:1), and the migrated spots were detected under UV illumination. The following R_F values were obtained: acenaphthene = 0.88; fluorene = 0.69; anthracene = 0.64; phenanthrene = 0.62; 9-methylanthracene = 0.57; pyrene = 0.52; fluoranthene = 0.51; benz[*a*]anthracene = 0.43; benzo[*a*]pyrene = 0.41; perylene = 0.28; chrysene = 0.22; benzo[*ghi*]perylene = 0.18; dibenz[*a*,*h*]-anthracene = 0.16; benzo[*e*]pyrene = 0.10; anthanthrene = 0.07.

The immobile–mobile phase technique was used by Stuchlik *et al.*,[5] who impregnated the paper with di-*n*-butyl ether and developed the chromatogram in a 0.05M solution of 1,3,7,9-tetramethyluric acid. The R_F values were: anthanthrene = 0.16; anthracene = 0.21; 2,3-benzofluorene = 0.06; 3,4-benzopyrene = 0.14; 1,2,5,6-dibenzanthracene = 0.05; 1,2,3,4-dibenzopyrene = 0.09; 1,2,4,5-dibenzopyrene = 0.11; 3,4,9,10-dibenzopyrene = 0.07; fluoranthene = 0.26; 20-methylcholanthrene = 0.08; pyrene = 0.31. Decahydronaphthalene was also used as an immobile phase, with the mobile phase being a 0.05M solution of 1-(2,3-dihydroxypropyl)theobromine. The R_F values for this system were: anthanthrene = 0.30; anthracene = 0.22; 2,3-benzofluorene = 0.05; 3,4-benzopyrene = 0.23; 1,2,5,6-dibenzanthracene = 0.12; 1,2,3,4-dibenzopyrene = 0.17; 1,2,4,5-dibenzopyrene = 0.20; 3,4,9,10-dibenzopyrene = 0.14; fluoranthene = 0.36; 20-methylcholanthrene = 0.10; pyrene = 0.37. In both cases the spots were detected under UV radiation.

2.1.3 Thin layer chromatography

Sawicki *et al.*[6] used thin layers of aluminum oxide or cellulose for the resolution of polycyclic hydrocarbons. Benzo[*a*]pyrene was used as the reference compound and migration distances were reported as R_B values.

Adsorbent: aluminum oxide

Developing solvent: pentane–ether (19:1)

R_B: phenanthrene = 1.13; anthracene = 1.14; fluoranthene = 1.09; chrysene = 1.10; pyrene = 1.25; triphenylene = 1.07; benz[*a*]-anthracene = 1.03; 11*H*-benzo[*b*]fluorene = 1.08; benzo[*e*]pyrene = 1.00; anthanthrene = 0.71; benzo[*ghi*]perylene = 0.89; dibenz[*ah*]-anthracene = 0.74; naphtho[1,2,3,4-*def*]chrysene = 0.78; benzo[*rst*]-pentaphene = 0.68; coronene = 0.46; benzo[*a*]coronene = 0.10; dibenzo[*hrst*]pentaphene = 0.12.

Adsorbent: cellulose

Developing solvent: formamide–water (1:1)

R_B: phenanthrene = 1.99; anthracene = 1.99; fluoranthene 1.89; chrysene = 1.75; pyrene = 1.72; triphenylene = 1.49; benz[*a*]-anthracene = 1.47; 11*H*-benzo[*b*]fluorene = 1.33; benzo[*e*]pyrene = 1.16; perylene = 1.14; benzo[*k*]fluoranthene = 1.03; benzo[*a*]-pyrene = 1.00; anthanthrene = 0.70; benzo[*ghi*]perylene = 0.69; dibenz[*ah*]anthracene = 0.66; naphtho[1,2,3,4-*def*]chrysene = 0.48; benzo[*rst*]pentaphene = 0.45; coronene = 0.37; benzo[*a*]-coronene = 0.15; dibenzo[*hrst*]pentaphene = 0.14.

Adsorbent: acetylated cellulose

Developing solvent: methanol–toluene–water (17:4:4)

R_B: phenanthrene = 3.74; anthracene = 3.33; fluoranthene = 2.92; pyrene = 3.16; benz[*a*]anthracene = 2.70; 11*H*-benzo[*b*]-fluorene = 3.54; benzo[*e*]pyrene = 2.94; perylene = 2.86; benzo[*k*]-fluoranthene = 2.40; benzo[*a*]pyrene = 1.00; anthanthrene = 2.17; benzo[*ghi*]perylene = 3.04; dibenz[*a,h*]anthracene = 2.92; naphtho[1,2,3,4-*def*]chrysene = 1.85; benzo[*rst*]pentaphene = 2.41; coronene = 2.87; benzo[*a*]coronene = 2.48; dibenzo[*h,rst*]pentaphene = 2.35.

In all these cases the migrated spots were detected by their fluorescence in UV radiation.

Silica gel was used as adsorbent by Ross[7] and developed in a saturated chamber with cyclohexane. Migrated spots were detected by reaction with 2,6-dichloroquinone chloroimide. The following compounds were chromatographed: chrysene, $R_F = 0.15$; pyrene, $R_F = 0.20$; anthracene, $R_F = 0.20$; phenanthrene, $R_F = 0.25$; naphthalene, $R_F = 0.20$; fluoranthene, $R_F = 0.20$.

White and Howard[8] used acetylated cellulose as the thin layer adsorbent and developed the chromatogram in ethanol, toluene and water (17:4:4). The spots were detected by their fluorescence under UV illumination. The following R_F values were obtained: benzo[*a*]pyrene = 0.23; 7,12-dimethylbenz[*a*]anthracene = 0.50; 4-methylpyrene = 0.70; 5,6-dimethylchrysene = 0.42; 3-methylcholanthrene = 0.57; anthracene = 0.57; 12-methylbenz[*a*]-anthracene = 0.34; 7,8-dimethylbenz[*a*]anthracene = 0.42; benzo[*c*]phenanthrene = 0.54; 1,2-dihydrobenz[*e*]aceanthrylene = 0.39; cholanthrene = 0.50;

5-methylchrysene = 0.39; pyrene = 0.62; fluoranthene = 0.60; 7-methylbenz-[*a*]anthracene = 0.45; 4-methylbenzo[*a*]pyrene = 0.37; benz[*a*]anthracene = 0.42; chrysene = 0.35; benzo[*k*]fluoranthene = 0.40; benzo[*e*]pyrene = 0.56; anthanthrene = 0.32; perylene = 0.47; benzo[*ghi*]perylene = 0.51; dibenz-[*ah*]anthracene = 0.48; dibenzo[*al*]pyrene = 0.24; dibenzo[*ai*]phenanth-rene = 0.57; dibenzo[*ae*]pyrene = 0.38; dibenz[*bi*]anthracene = 0.54.

Aza-heterocyclic compounds were resolved with cellulose and alumina layers.[9] Cellulose was used in conjunction with two developing solvent mixtures. The first consisted of dimethylformamide and water (35:65) and gave the following R_F values: 7-phenyldibenz[*ch*]acridine = 0.0; 8,12-dimethylbenz[*a*]acridine = 0.20; 8,10-dimethylbenz[*c*]acridine = 0.13; 7-methylbenz[*c*]acridine = 0.20; benz[*c*]acridine = 0.26; benzo[*h*]quinoline = 0.71; 7,10-dimethylbenz[*c*]acridine = 0.10; dibenz[*a,h*]acridine = 0.06; pyrenoline = 0.20; indeno[1,2,3-*ij*]isoquinoline = 0.64; acridine = 0.74; 3-methylbenzo[*f*]quinoline = 0.75; phenanthridine = 0.72; benzo[*f*]quinoline = 0.76; acenaphtho-[12-*b*]pyridine = 0.60; benzo[*lmn*]phenanthridine = 0.57; benz[*a*]acridine = 0.46; 12-methylbenz[*a*]acridine = 0.41; 9,12-dimethylbenz[*a*]acridine = 0.28; dibenz[*aj*]acridine = 0.22.

The second solvent mixture, composed of acetic acid and water (3:7) gave the following R_F values: 8,12-dimethylbenz[*a*]acridine = 0.62; 8,10-dimethylbenz[*c*]acridine = 0.40; 7-methylbenz[*c*]acridine = 0.61; benz[*c*]acridine = 0.55; benzo[*h*]quinoline = 0.77; 7,10-dimethylbenz[*c*]acridine = 0.53; 7,9-dimethylbenz[*c*]acridine = 0.46; pyrenoline = 0.26; indeno[1,2,3-*ij*]isoquinoline = 0.71; acridine = 0.83; 3-methylbenzo[*f*]quinoline = 0.79; phenanthridine = 0.78; benzo[*f*]quinoline = 0.79; naphtho[1,2-*b*]pyrine = 0.75; benzo[*lmn*]phenanthridine = 0.71; benz[*a*]acridine = 0.53; 12-methylbenz[*a*]acridine = 0.53; 12-methylbenz[*a*]acridine = 0.61; 9,12-dimethylbenz[*a*]acridine = 0.49; 14-phenyldibenz[*aj*]acridine = 0.12.

When alumina was used as the adsorbent it was developed with petroleum ether and ethyl ether (19:1). The following R_F values were obtained: 7-phenyldibenz[*c,h*]acridine = 0.70; 8,12-dimethylbenz[*a*]acridine = 0.60; 8,10-dimethylbenz[*c*]acridine = 0.54; 7-methylbenz[*c*]acridine = 0.53; benz[*c*]-acridine = 0.43; benzo[*h*]quinoline = 0.51; 7,10-dimethylbenz[*c*]acridine = 0.46; 7,9-dimethylbenz[*c*]acridine = 0.45; dibenz[*a,h*]acridine = 0.26; pyrenoline = 0.22; indeno[1,2,3-*ij*]isoquinoline = 0.19; acridine = 0.18; 3-methylbenzo[*f*]quinoline = 0.18; phenanthridine = 0.16; benzo[*f*]quinoline = 0.14; acenaphtho[1,2-*b*]pyridine = 0.13; benzo[*lmn*]phenanthridine = 0.12; benz[*a*]-acridine = 0.12; 12-methylbenz[*a*]acridine = 0.11; 9,12-dimethylbenz[*a*]acridine = 0.11; 14-phenyldibenz[*aj*]acridine = 0.08; dibenz[*aj*]acridine = 0.04.

Sawicki *et al.*[10] resolved aza-heterocyclic compounds on alumina with pentane and nitrobenzene (9:1). The spots were made visible by treating the layer with trifluoracetic acid and observing the fluorescence under UV radiation. The R_F values were: 7-phenyldibenz[*ch*]acridine = 0.94; benz[*c*]-acridine = 0.92; pyrenoline = 0.45; acridine = 0.37; benzo[*h*]quinoline =

0.35; benzo[*f*]quinoline = 0.30; 14-phenyldibenz[*aj*]acridine = 0.28; indeno[1,2,3-*ij*]isoquinoline = 0.26; benz[*a*]acridine = 0.23; benzo[*lmn*]-phenanthridine = 0.15.

Carbazole and polycyclic carbazoles can be separated on aluminum oxide layers in a variety of solvent systems.[11] The solvent mixture of pentane and chloroform (3:2) gave the following R_F values: carbazole = 0.50; 11*H*-benzo[*a*]carbazole = 0.46; 5*H*-benzo[*b*]carbazole = 0.39; 7*H*-benzo[*c*]carbazole = 0.37; 4*H*-benzo[*def*]carbazole = 0.49; 7*H*-dibenzo[*cg*]carbazole = 0.30. In most of these cases fluorescence observed under black light UV illumination was used to detect the polycyclic compounds. Little experimental work has been directed towards finding other reagents to give specific color reactions for the different types of compounds. So, whereas this detection method is reliable for standards, it can be subject to many interferences in natural samples.

2.2 WATER POLLUTION

Water, as it exists in the biosphere, occurs in four main forms; as groundwater, as freshwater on the surface, in the oceans and as vapor in the atmosphere. Chemical pollutants in natural water sources can be toxic to man and to organisms on whose existence man's survival depends. These pollutants can be converted to other forms by chemical and biochemical reactions.

Let us consider, for example, a factory on a river which discharges phenols into the water. If, further down the river there is a city which chlorinates its water supply, this treatment will transform the phenols into chlorinated compounds which are a greater potential hazard than the original chemicals.[3]

As our reliance on nuclear power for domestic purposes continues to increase, the prime concern of environmentalists appears to be the danger of radiation and thermal pollution. The operation of nuclear plants places a further demand upon natural water sources for efficient and economical cooling. The water used for cooling is treated with chemicals which are added to prevent the build-up of slime-producing organisms, and to prevent scale and corrosion, and the possibility always exists that these chemicals may find their way into primary water sources.

Surfactants are widely used for cleaning purposes, for the preparation of foods, for the sizing and treatment of cloth and other fibrous materials, for flotation mining processes and for many other applications, from which they may reach natural water sources. There can also be natural contamination of waters with cations and anions that could be detrimental to the health of humans and animals. The sampling of waters and the identification of pollutants has been described by Goulden.[3]

2.3 SURFACTANTS AND DETERGENT CHEMICALS IN WATER

A method for extracting surfactants from sewage was developed by

Patterson.[12] A 250 ml sewage sample was transferred to a separating funnel and diluted with 250 ml of water containing 150 g of dissolved magnesium sulfate. This solution was extracted with four separate 50 ml volumes of chloroform. Any emulsions formed were broken down by filtering through Whatman phase separation filter paper where the water stayed on the filter paper and the chloroform filtered through. The 200 ml volume of chloroform was washed with 200 ml of 1N hydrochloric acid and 50 g of sodium chloride; only the chloroform was drawn off into a clean separating funnel, and if necessary filtered through phase separation paper. The chloroform solution was then extracted with 200 ml of 1N sodium hydroxide and 50 g of sodium chloride, and drawn off into a suitable container. It was evaporated down to about 2 ml and transferred to a centrifuge tube, evaporated just to dryness and a further 0.5 ml of chloroform added. The sample could then be purified further, or used for chromatography directly.

Alumina[13] has proven to be very effective for cleaning samples containing surfactants. An alumina, such as Woelm Neutral, was activated for use by heating at 800 °C for 3 h, cooling in a desiccator and then storing it under desiccating conditions until ready for use. The column was packed with 40 g of adsorbent made into a slurry with chloroform. The surfactant extract was dissolved in chloroform and introduced into the column, which was first eluted with chloroform to displace the long chain tertiary amines, a trace of amine oxide and *N*-hydroxyalkyl methyl tauride with 60% of the Hyamine 1622. The second eluent, ethanol, displaced 90% of the amine oxide, a trace of the *N*-hydroxyalkyl methyl tauride, pyridine sulfobetaine and alkyl quaternary sulfobetaine, and the rest of the Hyamine 1622. The third eluent, 90% aqueous ethanol, removed traces of amine oxide, 80% of the *N*-hydroxyalkyl methyl tauride, some of the alkene sulfonate and the alkyl benzene sulfonates. Increasing the water content to 50% eluted the remainder of the sulfonates.

Konig[14] used strong and weak ion-exchange resins for separating anionic detergents into sulfated–sulfonated surfactants and carboxylic type surfactants. A column 10 cm long and 2 cm in diameter was filled with Dowex 50WX4 or 1X2 ion-exchange resins. The Dowex 50WX4 was activated by equilibration with a 10% sodium chloride solution for 24 h. The Dowex 1X2 resin was equilibrated with 2N sodium hydroxide or 2N sodium chloride. The columns were packed and then preeluted with 100 ml of ethanol. The samples were introduced into the column in 96% ethanol and the column was prewashed with ethanol. The hydrogen form of the Dowex 50WX4 was eluted with concentrated hydrochloric acid and ethanol (1:1). Surfactants of the aminocarboxylic acid type, betaine compounds and imidazole derivatives were eluted; sulfates and sulfonates were recovered by eluting the Cl^- form of the Dowex 1X2 with 2% sodium chloride and isopropanol (1:1); the sodium salts of fatty acids and aminocarboxylic acids were eluted from the OH^- form of Dowex 1X2 with 2% sodium hydroxide and isopropanol (1:1).

2.3.1 Paper chromatography

A general identification procedure was developed by Drewry.[15] Whatman No. 1 paper was prewashed, first with 0.2N hydrochloric acid solution and then four times with distilled water and finally with 15% aqueous ethanol. The paper was dried at room temperature. *t*-Butanol, ammonium hydroxide and water (100:6:6) proved to be a suitable solvent system. The papers were developed overnight for 18 h, then dried in an oven at 80 °C; the migrated spots were detected by spraying with a 0.05% ethanolic solution of Pinacryptol yellow and then observing the fluorescence under UV illumination.

Nonionic detergents gave an R_F of 0.9–0.95 just below the solvent front and were visible as pale blue spots. Amine oxides were visible as pale blue spots with an R_F of 0.85. Anionic detergents were usually observed as elongated brilliant orange spots with an R_F of 0.75. At the same R_F, sulfonated alkylphenol ethoxylates were visible as violet spots, while soaps formed pale yellow spots which eventually faded. Cationic detergents gave an R_F of 0.70; the spots were elongated and pale blue. Toluene sulfonates formed reddish-orange spots with an R_F of 0.25, and xylene sulfonates gave yellow spots with an R_F of 0.32. Ethanolamines and cations were observed as faint grey-violet spots near the origin.

Bromcresol green was used with the above solvent systems as a chromogenic reagent, and it detected alkaline substances as blue spots; 0.04 g were dissolved in ethanol and then titrated with dilute ammonia solution until the resulting solution just turned green. The following R_F values were observed: amine oxides = 0.85; cationic detergents = 0.70; triethanolamine = 0.57; monoethanolamine = 0.43; monoisopropanolamine = 0.50; metals = 0.0. A cobaltothiocyanate chromogenic reagent was also used. It formed bluish-green spots with polyethoxylates, amine oxides and cationic detergents. The reagent was prepared by dissolving 1.4 g of cobalt nitrate and 85 g of ammonium thiocyanate in water and diluting to 500 ml. It was diluted with an equal volume of ethanol and sprayed onto the chromatograms.

Another developing solvent system used was ethyl acetate, methanol and ammonium hydroxide (60:15:5). The ascension time for this solvent was about three hours. Anionic compounds, $R_F = 0.70$; nonionic compounds, $R_F = 0.95$; soaps, $R_F = 0.70$; quaternary cationic compounds, $R_F = 0.60$; *p*-toluene sulfonate, $R_F = 0.15$; xylene sulfonate, $R_F = 0.22$; urea, $R_F = 0.24$.

With this system ninhydrin could be used as a reagent for detecting primary amines and other nitrogenous compounds; 0.2 g was dissolved in 95 ml of methanol and 5 ml of 2N acetic acid was added. The chromatogram was sprayed with the reagent and heated at 80 °C for 5 min. Primary amines gave purple spots, but if the chromatograms were heated at 120 °C for 10 min the spots already formed deepened in color and other nitrogenous compounds present gave purple or blue spots.

2.3.2 Thin layer chromatography

Anionic surfactants were chromatographed by the methods of Bey[16] or Mangold and Kammereck.[17] Bey used Silica Gel G as the adsorbent and a developing solution of *n*-propanol, chloroform, methanol and 10N ammonium hydroxide (10:10:5:2). The following R_F values were obtained: Seifenmehl = 0.4; Medialan KA = 0.37; Lamepon A = 0.26; Lamepon S = 0.3, 0.81; Lamepon A44 = 0.29; Texapon = 0.5; Texapon THR = 0.49; Lensodel AB = 0.5; Texapon N25 = 0.3; Fenopon CO436 = 0.5; Somepon ML = 0.49; Avirol DKM = 0.45; Levapon = 0.49; Marlon A350 = 0.5; Nansa HS = 0.80, 0.45; Eltesol ST 90 = 0.4; Cumulsulfonat = 0.42; Nekal BX = 0.54, 0.67; Triton X-200 = 0.6; Hostapon KA = 0.55; Hostapon KTW = 0.3, 0.53; Manoxol OT = 0.6; Steinapol SBF = 0.05; Steinazid SBU = 0.08; Hostaphot KW200 = 0.59, 0.85, 0.89; Hostaphot K280 = 0.69, 0.85. Migrated spots were detected by spraying with a 0.05% ethanolic solution of pinacryptol yellow and their fluorescence observed under UV illumination.

Mangold and Kammereck[17] also used Silica Gel G and a developing solvent containing 97 volumes of chloroform equilibrated with 1N aqueous ammonia added to 3 volumes of methanol. The spots were made visible under UV illumination after spraying the chromatogram with 0.2% ethanolic dichlorofluorescein solution. They found the following R_F values: Seifenmehl = 0.85; Medialan KA = 0.66; Lamepon A = 0.0; Lamepon S = 0.0; Lamepon A44 = 0.0; Texapon K12 = 0.25; Texapon TH = 0.22; Lensodel AB6 = 0.21; Texapon N25 = 0.25; Fenopon CO436 = 0.23; Somepon ML = 0.16; Avirol DKM = 0.17; Levapon T = 0.24; Marlon A350 = 0.25; Nansa HS80 = 0.22; Etesol ST90 = 0.16; Cumulsulfonat = 0.16; Nekal BX = 0.2; Triton X-200 = 0.39, 0.5, 0.65; Hostapon KA = 0.19; Hostapon KTW = 0.13; Manoxol OT = 0.33; Steinapol SBF = 0.85; Steinapol SBZ = 0.75; Steinazid SBU = 0.0; Hostaphot KW200 = 0.32, 0.82; Hostaphot KO280 = 0.49, 0.79, 0.82, 0.93.

Table 2.2 lists the names of the surfactants that were analyzed and their chemical constituents.

2.4 PHENOLS AND RELATED COMPOUNDS

Churacek *et al.*[18] utilized both paper and thin layer chromatography in order to identify many phenolic compounds. The phenols were converted to their *p*-(*N*,*N*-dimethylaminobenzene)-*p*′-azobenzoates and then chromatographed. The extracted and cleaned sample was reacted with 150 mg of *p*-(*N*,*N*-dimethylaminobenzene)-*p*′-azobenzyl chloride in a solution of pyridine (205 ml), heated, and refluxed for 1 h. The reaction solution was then cooled and 2 ml of 5% sodium carbonate solution added, after which the solution was filtered and the filtrate used for spotting on paper or thin layer chromatograms.

TABLE 2.2
Composition of commercial surfactants

Trade name	Description
Seifenmehl	Sodium stearate
Medialan KA	Sodium salt of coconut oil fatty acids reacted with sarcosinic acid
Lamepon A	Oleic acid chloride reacted with sodium salts of amino acids
Lamepon S	Amino acids from albumin hydrolysate reacted with potassium salts of vegetable oil fatty acids
Lamepon A44	Albumin hydrolysate reacted with alkylsulfonyl chloride
Texapon K12	Sodium lauryl sulfate
Texapon TH	Triethanolamine lauryl sulfate
Lensodel AB6	Secondary sodium alkyl sulfate
Texapon N25	Sodium lauryl ethoxy sulfate
Fenepon CO436	Ammonium salt of alkyl phenylpolyglycol ethoxy sulfate
Somepon ML	Sodium salt of fatty acid alkanolamide sulfates
Avirol DKM	Sodium salt of sulfated castor oil
Levapon T	Sodium paraffin sulfonate
Marlon A350	Sodium dodecylbenzyl sulfonate
Nansa HS/80 Schuppen	Sodium tetrapropylbenzyl sulfonate
Eltesol ST90	Sodium toluyl sulfonate
Cumulsulfonat	Sodium cumenyl sulfonate
Nekal BX	Sodium alkylnaphthalene sulfonate
Triton X-200	Sodium octyl phenylpolyglycol ethoxy sulfonate
Hostapon KA	Sodium salt of coconut oil fatty acid isothionates
Hostapon KTWA	Sodium salt of coconut oil fatty acid taurides
Manoxol OT	Sodium salt of dioctylsulfosuccinate
Steinapol SBFA30	Sodium salt of the half-ester of laurylethoxy sulfosuccinate
Steinapol SBZ	Sodium salt of coconut oil alkylamide ethoxy sulfosuccinate
Stinazid SBU185	Sodium salt of undecylinic acid reacted with monoethanolamido-sulfosuccinate
Hostaphot KW200	Higher wax alcoholic esters of secondary *O*-phosphoric acid
Hostaphot KO280	Secondary esters of *O*-phosphoric acid with oleicoctaglycolic esters

2.4.1 Paper chromatography

Whatman No. 1 chromatographic paper was used in conjunction with several immobile–mobile phase systems.

Immobile phase: 30% *N,N*-dimethylformamide

Mobile phase: hexane

R_F: phenol = 0.41; *o*-cresol = 0.49; *m*-cresol = 0.53; *o*-xylenol = 0.56; *o*-nitrophenol = 0.19; *m*-nitrophenol = 0.23; *p*-methoxyphenol = 0.24; *o*-chlorophenol = 0.40; 1-naphthol = 0.45; *o*-*sec*-butylphenol = 0.76; *o*-octyl-*o*-cresol = 0.62; *p*-*tert*-butyl-*o*-cresol = 0.73.

Immobile phase: 40% *N,N*-dimethylformamide

Mobile phase: hexane–xylene (10:1)

R_F: phenol = 0.48; *o*-cresol = 0.52; *m*-cresol = 0.53; *o*-xylenol = 0.58; *o*-nitrophenol = 0.11; *p*-methoxyphenol = 0.22; 1-naphthol = 0.47; cetylphenol = 0.63; *p*-*tert*-butyl-*o*-cresol = 0.66.

Immobile phase: 40% *N*,*N*-dimethylformamide
Mobile phase: hexane–xylene (10:1)
R_F: phenol = 0.38; *o*-cresol = 0.51; *m*-cresol = 0.54; *o*-xylenol = 0.70; *m*-nitrophenol = 0.16; *p*-methoxyphenol = 0.19; *o*-chlorophenol = 0.30; 2-methyl-6-octylphenol = 0.82; *p*-*tert*-butyl-*o*-cresol = 0.85; 4,6-di-*tert*-butyl-*o*-cresol = 0.61.

Immobile phase: 10% paraffin oil
Mobile phase: dimethylformamide–water–methanol (8:2:3)
Octylphenol, R_F = 0.20; nonylphenol, R_F = 0.13; dodecylphenol, R_F = 0.08; *p*-*tert*-octylphenol, R_F = 0.19; *o*-*sec*-butylphenol, R_F = 0.50; *o*-octyl-*o*-cresol, R_F = 0.17; *o*-*tert*-butyl-*p*-cresol, R_F = 0.46; *p*-*tert*-butyl-*o*-cresol, R_F = 0.40.

Immobile phase: 10% paraffin oil
Mobile phase: dimethylformamide–butanol–water (80:5:15)
Octylphenol, R_F = 0.51; nonylphenol, R_F = 0.39; dodecylphenol, R_F = 0.20; cetylphenol, R_F = 0.03; *p*-*tert*-butyl-*o*-cresol, R_F = 0.68.

2.4.2 Thin layer chromatography

Adsorbent: Silica Gel CH
Developing solvent: cyclohexane–methyl ethyl ketone (7:3)
R_F: phenol = 0.66; *o*-cresol = 0.67; *m*-cresol = 0.69; *p*-methoxyphenol = 0.64; *o*-chlorophenol = 0.75; *o*-nitrophenol = 0.54; *m*-nitrophenol = 0.63; octylphenol = 0.79; *o*-xylenol = 0.55; nonylphenol = 0.76; dodecylphenol = 0.80; cetylphenol = 0.80; *o*-*sec*-butylphenol = 0.69; *o*-octyl-*o*-cresol = 0.75; *o*-*tert*-butyl-*o*-cresol = 0.70; *p*-*tert*-butyl-*o*-cresol = 0.73; *p*-*tert*-octylphenol = 0.70.

Adsorbent: Silica Gel CH
Developing solvent: cyclohexane–ethyl acetate (4:1)
R_F: phenol = 0.53; *o*-cresol = 0.50; *p*-cresol = 0.49; *p*-methoxyphenol = 0.39; *o*-nitrophenol = 0.32; *m*-nitrophenol = 0.40; *p*-nitrophenol = 0.47; *o*-xylenol = 0.51; nonylphenol = 0.56; *o*-*sec*-butylphenol = 0.51; *o*-octyl-*o*-cresol = 0.55; *o*-*tert*-butyl-*o*-cresol = 0.56; *p*-*tert*-butyl-*o*-cresol = 0.56; *p*-*tert*-octylphenol = 0.57; *o*-*tert*-butyl-*p*-cresol = 0.52; 4,6-di-*tert*-butyl-*o*-cresol = 0.48.

Adsorbent: Silica Gel G
Developing solvent: hexane–ethyl acetate (4:1)
R_F: phenol = 0.33; *o*-cresol = 0.31; *m*-cresol = 0.45; *p*-methoxyphenol = 0.15; *o*-chlorophenol = 0.25; *m*-nitrophenol = 0.20; nonylphenol = 0.50; *o*-*sec*-butylphenol = 0.40; *o*-octyl-*o*-cresol = 0.45; *o*-*tert*-butyl-*o*-cresol = 0.47; *p*-*tert*-octylphenol = 0.48; 1-naphthol = 0.23.

Adsorbent: Silica Gel G
Developing solvent: benzene–chloroform (1:1)

R_F: phenol = 0.42; *o*-cresol = 0.38; *m*-cresol = 0.50; *p*-methoxyphenol = 0.25; *o*-chlorophenol = 0.50; *o*-nitrophenol = 0.32; nonylphenol = 0.59; dodecylphenol = 0.69; cetylphenol = 0.70; 1-naphthol = 0.38.

Adsorbent: Silica Gel G

Developing solvent: chloroform

Phenol, R_F = 0.55; *m*-cresol, R_F = 0.75; *p*-methoxyphenol, R_F = 0.46; *o*-chlorophenol, R_F = 0.80; *m*-nitrophenol, R_F = 0.49; dodecylphenol, R_F = 0.80; cetylphenol, R_F = 0.83; 1-naphthol, R_F 0.59.

Since the derivatives of these compounds are easy to chromatograph and they are reddish-orange in color they have a built-in indicator which makes them visible without any further treatment.

Smith and Lichtenberg[19] were able to separate some chlorinated phenols using Silica Gel G and developing with chloroform. The R_F values found were: *o*-chlorophenol = 0.56; *m*-chlorophenol = 0.20; *p*-chlorophenol = 0.20; 2,3-dichlorophenol = 0.55; 2,4-dichlorophenol = 0.55; 2,5-dichlorophenol = 0.58; 2,6-dichlorophenol = 0.69; 3,4-dichlorophenol = 0.21; 2,4,5-trichlorophenol = 0.50; 2,4,6-trichlorophenol = 0.67; 4-chloro-2-nitrophenol = 0.67.

Zigler and Philips[20] developed the following procedure for separating some chlorinated phenols by thin layer chromatography. A liter sample of water was acidified with phosphoric acid and then extracted with four separate 100 ml volumes of petroleum ether. The extracts were combined and evaporated down to 5–10 ml. The resulting extract was dried with sodium sulfate and evaporated down to about 0.1 ml in a centrifuge tube. This volume was then spotted onto an Aluminum Oxide G plate with a calcium sulfate binder. A sandwich chamber was used both for development and clean-up. After the sample was spotted and the plate placed in the sandwich chamber, it was developed 3 inches in benzene, then dried and the process repeated three or four times. This chromatographed some of the interferences away from the origin.

The chlorophenols were then resolved by development in a solution of 6 ml of 1N sodium hydroxide and 94 ml of acetone The plate was then dried and sprayed with silver nitrate solution (0.5 g $AgNO_3$ in 5 ml water + 100 ml 2-phenoxyethanol diluted to 1 liter with acetone and 3 drops of 30% hydrogen peroxide). The plate was dried again and then exposed to high intensity UV radiation for 15 min, which rendered the spots visible (light or dark brown). The following R_F values were obtained: 3-chlorophenol = 0.94; 2,4-dichlorophenol = 0.71; 2,4,5-trichlorophenol = 0.62; 2,4,6-trichlorophenol = 0.42; pentachlorophenol = 0.09.

Aly[21] used a derivatization technique to separate phenols from chlorophenols. He took a 100 ml water sample and adjusted the pH to 8.0 by adding 5 ml of phosphate buffer and 5 ml of 0.5N ammonium hydroxide solution. Two ml of 4-aminoantipyrine were then added, plus 1 ml of potassium ferricyanide solution, the mixture being thoroughly shaken after each addition. A red color developed when the solution was allowed to react for 15 min. This red dye was then extracted with two 5 ml portions of chloroform and dried with anhydrous magnesium sulfate. The solvent was concentrated to about 0.5 ml and spotted onto silica gel plates. The red color served as a built-in chromogenic reagent.

Three solvent systems were used:

(1) Methylene chloride–acetylacetone (70:30)

R_F: phenol = 0.40; 2-chlorophenol = 0.64; 3-chlorophenol = 0.52; 4-chlorophenol = 0.40; 2,4-dichlorophenol = 0.64; 2,6-dichlorophenol = 0.90; 2,4,6-trichlorophenol = 0.90.

(2) Chloroform–ethyl acetate (80:20)

R_F: phenol = 0.15; 2-chlorophenol = 0.29; 3-chlorophenol = 0.32; 4-chlorophenol = 0.15; 2,4-dichlorophenol = 0.28; 2,6-dichlorophenol = 0.53; 2,4,6-trichlorophenol = 0.53.

(3) Ethyl acetate–hexane–acetic acid (70:20:10)

R_F: phenol = 0.35; 2-chlorophenol = 0.45; 3-chlorophenol = 0.52; 4-chlorophenol = 0.35; 2,4-dichlorophenol = 0.45; 2,6-dichlorophenol = 0.53; 2,4,6-trichlorophenol = 0.53.

2.5 METALS IN WATER

Industrial effluents and domestic sewage have increased the concentration of cations and anions in water supplies used for domestic and agricultural purposes. Human populations are therefore exposed to larger amounts of these elements over longer periods of time than has been the case in the past. The long range toxic effects of such exposure are just beginning to be studied.

Cadmium is not an essential element, but it is present in the soil, in plants and in the food supply. It can enter our environment from mining and metallurgy, chemical industries, scrap metal treatment, electroplating, superphosphate fertilizers and cadmium-containing pesticides. It is known that cadmium can accumulate in some marine animals, but not much is known about its transformation in the environment.

Arsenic is not essential for living systems but is regularly found in human and other living tissues in trace amounts. It is at present under suspicion as being a carcinogen in certain forms.

Lead is a toxicant which is well known to be especially dangerous to children. It usually enters the environment from pigment manufacturing, lead smelting and refining, storage battery production and the combustion of fuels.

Mercury occurs naturally in the environment; it is used in medicine and is also a byproduct of mining, chlorine alkali plants, pulp and paper manufacture, plastics and electronic industries, and the use of drugs and agricultural chemicals. It is usually discharged into the environment in the form of inorganic or alkyl, alkoxy, aryl or other organic mercury compounds. It has recently been found that mercury can be converted into the highly toxic methyl mercury form by microorganisms. This form can thus enter the food chain.

Antimony can be released from industrial metal-processing and from safety matches. Beryllium in the environment originates from the production of the metal and its alloys, and can be highly toxic to the respiratory system. Nitrates and nitrites may originate from fertilizers.

Ordinarily, paper and thin layer chromatography are not associated with the analysis of metals. Spectroscopy and colorimetry are the usual analytical methods of choice but chromatography can also be quite valuable as an identification tool.[22]

2.5.1 Paper chromatography of cations

Kertes and Lederer[23] applied paper chromatography to the resolution of several cations.

Chromatographic paper: Whatman No. 1

Solvent system: butanol–hydrobromic acid–water (100:10:90). The butanol layer is used.

R_F: Cd = 0.80; Hg^{2+} = 1.0; Cu^{2+} = 0.10; Ag = 0.0; Pb = 0.40; As^{3+} = 0.54; Cr = 0.03; Th = 0.0; U^{6+} = 0.11; U^{4+} = 0.0 When 1 ml of hydrobromic acid was added to the butanol layer the following R_F values were obtained: Cd = 1; Hg – 1; Cu = 1; Pb = 0.60; As = 0.78; Cr = 0.06; Th = 0.03; U^{6+} = 0.21; U^{4+} = 0.2.

With the volume of hydrobromic acid added increased to 60 ml R_F: Pb = 0.65; Cr = 0.30; Th = 0.18; U^{6+} = 0.47; U^{4+} = 0.25.

Ascending chromatography was used, and a 24 h equilibration period in the solvent vapors was allowed before the chromatogram was developed.

Carboxymethyl cellulose paper in conjunction with a solvent system of *n*-butanol, 15% trichloracetic acid and 1-phenyl-1,3-butanedione (100:100:1) (V:V:W) was used by Elbeih and Abou-Elnaga.[24] The results were as follows: As^{3+}, R_F = 0.30; Ag, R_F = 0.40; Pb, R_F = 0.48; Th^{4+}, R_F = 0.59; Cd, R_F = 0.63; Co, R_F = 0.64; Hg^{+}, R_F = 0.70; Cu^{2+}, R_F = 0.74; Hg^{2+}, R_F = 0.83; U^{6+}, R_F = 0.94.

Grassini and Ossicini[25] resolved some anions with paper chromatography, using Whatman No. 1 paper with a developing solvent of isopropanol (A) in combination with a range of volumes of 1.5N ammonium hydroxide (B); the following R_F values were obtained:

Solvent (1) 80A:20B

Chloride = 0.32; selenite = 0.05; selenate = 0.08; nitrite = 0.43; arsenite = 0.11; arsenate = 0.01; oxalate = 0.03.

Solvent (2) 65A:35B

Chloride = 0.55; selenite = 0.26; selenate = 0.38; nitrite = 0.69; arsenite = 0.25; arsenate = 0.16; oxalate = 0.26.

Solvent (3) 55A:45B

Chloride = 0.68; selenite = 0.43; selenate = 0.51; nitrite = 0.62; arsenite = 0.33; arsenate = 0.34; oxalate = 0.44.

Solvent (4) 40A:60B

Chloride = 0.77; selenite = 0.65; selenate = 0.68; nitrite = 0.76; arsenite = 0.46; arsenate = 0.62; oxalate = 0.65.

2.5.2 Thin layer chromatography of cations

Some heavy metal cations were chromatographed on silica gel in the dithionate form which behaved as ready-made indicators since the salts are colored.[26]

Adsorbent: Silica Gel G

Solvent system (1) carbon tetrachloride–chloroform (20:4)

R_F: Pb = 0.39; Cu = 0.64; Hg = 0.74; Cd = 0.25; Zn = 0.73; Mn = 0.69; Ni = 0.38; Co = 0.50.

Solvent system (2) toluene

R_F: Pb = 0.37; Cu = 0.55; Hg = 0.74; Cd = 0.18; Zn = 0.70; Mn = 0.67; Ni = 0.35; Co = 0.38.

Solvent system (3) xylene

R_F: Pb = 0.39; Cu = 0.38; Hg = 0.58; Cd = 0.11; Zn = 0.63; Mn = 0.49; Ni = 0.18; Co = 0.20.

Solvent system (4) benzene

R_F: Pb = 0.43; Cu = 0.61; Hg = 0.76; Cd = 0.20; Zn = 0.69; Mn = 0.66; Ni = 0.38; Co = 0.48.

Heavy metal dithionate salts show the following colours: lead = red; copper = brown; mercury = orange; cadmium = reddish-orange; zinc = scarlet; manganese = crimson; nickel = greyish-violet; cobalt = violet.

Lederer and Rinalduzzi[27] used Polygram Polyamid II plastic precoated sheets to resolve many cations. The most satisfactory solvents were combinations of acetone (A) and 5N aqueous hydrochloric acid (B).

Solvent system (1) 90A:10B

R_F: Pr = 0.37; Hg = 0.31; Bi = 0.05; Cu = 0.33; Cd = 0.42; Au = 0.62; Sb = 0.58; Fe = 0.72; Tl = 0.72; Co = 0.38; Mn = 0.39; Ni = 0.10; UO_2 = 0.23; Zn = 0.34; Sn = 0.61; Pt = 0.06; Pd = 0.05; As = 0.61.

Solvent system (2) 80A:20B

R_F: Pb = 0.31; Hg = 0.27; Bi = 0.08; Cu = 0.35; Cd = 0.43; Au = 0.64; Sb = 0.42; Fe = 0.64; Tl = 0.70; Co = 0.28; Mn = 0.30; Ni = 0.22; UO_2 = 0.17; Zn = 0.35; Sn = 0.65; Pt = 0.04; Pd = 0.06; As = 0.64.

Solvent system (3) 60A:40B

R_F: Pb = 0.38; Hg = 0.13; Bi = 0.06; Cu = 0.41; Cd = 0.34; Au = 0.08; Sb = 0.17; Fe = 0.09; Co = 0.60; Mn = 0.84; Ni = 0.78;

$UO_2 = 0.07$; Zn = 0.25; Sn = 0.42; Pt = 0.03; Pd = 0.10; As = 0.73.

The samples were prepared by dissolving the metals in aqueous 3N hydrochloric acid, and the migrated spots were detected with hydrogen sulfide, ammonium sulfide or ammoniacal ethanolic 8-hydroxyquinoline.

REFERENCES

1. E. Sawicki, R. C. Corey, A. E. Dooley, J. B. Giselard, J. L. Menkman, R. E. Nelligan and L. A. Ripperton, *Health Lab. Sci.* **7**, pt 2, 31 (1970).
2. E. Sawicki, R. C. Corey, A. E. Dooley, J. B. Giselard, J. L. Menkman, R. E. Nelligan and L. A. Ripperton, *Health Lab. Sci.* **7**, pt 2, 45 (1970).
3. P. D. Goulden, *Environmental Pollution Analysis*, Heyden, London, 1978.
4. L. Dubois, A. Corkery and J. L. Monkman, *Int. J. Air Pollut.* **2**, 236 (1960).
5. M. Stuchlik, L. Krasnec and L. Csiba, *J. Chromatogr.* **30**, 543 (1967).
6. E. Sawicki, T. W. Stanley, W. C. Elbert and J. D. Pfaff, *Anal. Chem.* **36**, 497 (1964).
7. J. H. Ross, *Anal. Chem.* **60**, 2140 (1968).
8. R. H. White and J. W. Howard, *J. Chromatogr.* **29**, 108 (1967).
9. E. Sawicki, T. W. Stanley, J. D. Pfaff and W. C. Elbert, *Anal. Chim. Acta* **31**, 359 (1964).
10. E. Sawicki, W. C. Elbert and T. W. Stanley, *J. Chromatogr.* **27**, 170 (1965).
11. D. F. Bender, E. Sawicki and R. M. Wilson, *Anal. Chem.* **36**, 1011 (1964).
12. S. J. Patterson, *J. Inst. Sewage Purif.* 190 (1960).
13. S. J. Patterson, *J. Inst. Sewage Purif.* 378 (1960).
14. H. Konig, *Z. Anal. Chem.* **254**, 337 (1971).
15. J. Drewry, *Analyst* **88**, 225 (1963).
16. K. Bey, *Fette Seifen Anstrichm.* **67**, 217 (1965).
17. H. K. Mangold and R. Kammereck, *J. Am. Oil Chem.* **39**, 201 (1962).
18. J. Churacek, H. Pechova and V. Mares, *J. Chromatogr.* **67**, 97 (1972).
19. D. Smith and J. J. Lichtenberg, *ASTM Spec. Techn. Publ.* No. 448, **78** (1969).
20. M. G. Zigler and W. F. Philips, *Env. Sci. Technol.* **1**, 65 (1967).
21. O. M. Aly, *Water Res.* **2**, 587 (1968).
22. E. Berman, *Toxic Metals and their Analysis*, Heyden, London (1979).
23. S. Kertes and M. Lederer, *Anal. Chim. Acta* **15**, 543 (1956).
24. I. I. N. Elbeih and M. A. Abou-Elnaga, *Chem. Anal.* **56**, 99 (1967).
25. G. Grassini and L. Ossicini, *J. Chromatogr.* **7**, 351 (1962).
26. S. N. Tewari and N. Bhatt, *Mikrochim. Acta* 338 (1973).
27. M. Lederer and B. Rinalduzzi, *J. Chromatogr.* **68**, 237 (1972).

CHAPTER 3

PESTICIDES

All organisms in the biosphere compete with other organisms for survival. As part of its struggle to survive, the human species is primarily concerned with the utilization of natural resources in an efficient and nondamaging manner. Man requires food for nourishment, fiber for clothing, and a defense against pathogens and other organisms which may be dangerous to himself, or may compete for his resources. Pesticides are commonly used by human communities, as part of such a defense, to suppress pathogens and other competing organisms.

The earliest pesticides used were plant extracts such as pyrethrins and nicotine. Sulfur was used as a fungicide and other inorganic salts were also found to have insecticidal properties. As the field of organic chemistry has developed, the synthesis of new molecules has also accelerated so that chemical pesticides may now be made synthetically. Perhaps the most widely publicized of these new synthetic chemicals is DDT, which has proved an effective and inexpensive insecticide in the past and is still widely used to control malaria.

Although the use of pesticides has helped to control insect-borne diseases and to increase food production, workers such as Rachel Carson in *Silent Spring*[1] have highlighted the latent dangers involved in the indiscriminate use of these chemicals. This concern has now extended to all forms of potential environmental pollution.

The awareness of these potential dangers has compelled many governments to set up laws governing the use of pesticides. In the United States, the Environmental Protection Agency is committed to review the proposed use of all pesticides and to set tolerance levels for them in consumable commodities. The defined tolerance levels are not absolute toxicity levels; they fall below toxic levels and act as a safeguard as well as a legal limit to the total amount of residue that may be consumed. The Food and Drug Administration section of the Department of Health, Education and Welfare is assigned by law to enforce these tolerance levels, and one of their initial screening tests for detecting organochlorine insecticides at the microgram level utilizes paper chromatography. Indeed, both paper and thin layer chromatography are in use in countries all over the world for the detection of pesticide residues in food products and in the environment.

Insecticides are some of the most extensively applied pesticides, and are used to protect growing and stored crops from insect damage and to protect human and animal populations from disease-carrying insects. They may be divided into three main types: organochlorine compounds, organophosphorus compounds, and carbamates. Currently, a new series of synthetic pyrethroids are also being manufactured and tested for their effectiveness.

Matrices that need to be monitored for pesticide residues include food, vegetation, domestic and feral animals, water, soil, air and, when possible, human tissues.

The development of analytical methods for pesticides is complicated by the fact that in many cases the residue is no longer in the original form. The transformation usually occurs when the pesticide enters a living organism. It is foreign to the organism's biochemical processes, so in order to combat any possible detrimental effects to itself the organisms alters the molecule to a form which may be eliminated in the waste product cycle, or utilized in normal metabolic processes or stored in a nontoxic state.

In some cases the altered pesticide may prove to be more toxic than the original form, thus Parathion becomes converted to Paraoxon which is much more toxic as a cholinesterase inhibitor. Bees also convert Malathion to Malaoxon which is much more toxic to them than the original molecule.

The processes of detoxification are of four major types: oxidation, reduction, hydrolysis and synthesis. These biological reactions can form residues with solubilities ranging from lipophilic to hydrophilic, with many having dual solubilities which could be referred to as 'amphiphilic'. It is these dual solubility molecules that present the greatest difficulties during the isolation and purification stages in preparation for paper or thin layer chromatography.

3.1 EXTRACTION AND CLEAN-UP

The first important type of residue-containing matrix to be considered for analysis is food, and because the initial pesticides are generally lipophilic it may be considered that a fat-dissolving agent such as a saturated hydrocarbon or benzene is appropriate for extracting the residue. A search of the literature indicates that the research performed by von Müller *et al.*[2] in 1957 has set the basic approach for later methods to follow.

This group was trying to develop a general procedure for some organochlorine and organophosphorus insecticides using paper chromatography as the final detection step. Their general approach was to apply the Soxhlet extraction procedure for isolating the pesticide residue, using pentane which was found to be a suitable solvent. The food product containing the residue was dried and extracted for 6–8 h, and the resulting pentane extract was divided into three portions: one to be used to identify any organochlorine insecticides, the second to identify any organophosphorus insecticides and the third aliquot was kept in reserve.

The pentane solvent was evaporated on a steam bath. The flask containing the residue was cooled to room temperature and the residue then taken up with several volumes of acetonitrile and filtered into another flask (the acetonitrile-insoluble material remained on the filter). The acetonitrile was then distilled at 105 °C, and the residual material was taken up with pentane, after which it was ready for further purification on an alumina adsorption column. (The alumina column was prepared by making a pentane slurry and packing the adsorbent into the column.) The sample was placed onto the column with pentane and then eluted with ethyl ether.

Samples containing large amounts of wax were concentrated to a small volume in pentane and placed in a dry ice–isopropanol bath which froze out the waxes. The solution was then filtered and the container rinsed several times with cooled pentane. This pentane solution was then placed on the alumina column.

The procedure used for the organophosphorus resides was an adaptation of the Jones and Riddick partitioning technique.[3] The acetonitrile extract described earlier in this section was placed in a separating funnel and partitioned with pentane to remove lipophilic pigments and other materials. The volume of acetonitrile was kept greater than that of the pentane throughout the partitioning process (at least 2:1). An alumina column was prepared by slurrying with acetonitrile and pouring into the column. The sample was dissolved in acetonitrile and transferred to the column which was eluted with the same solvent.

In order to facilitate spotting on paper, it was desirable to have the residue in a volatile solvent such as pentane. To this end, the residue was partitioned back into pentane by watering out the acetonitrile with an acidic solution of water. In practice, the acetonitrile was placed in a separating funnel with a small volume of concentrated hydrochloric acid in water, and then a volume of pentane larger than the volume of acetonitrile was added. The aqueous layer was drawn off after a first partitioning and then partitioned a second time with pentane. The pentane layers were combined, dried over sodium sulfate and concentrated for spotting on a paper chromatogram.

In 1959, Klein[4] compared the relative efficiencies of Soxhlet extraction, tumbling, stripping and mechanical homogenization using benzene and benzene–isopropanol as the extracting solvents. The residues were labelled with ^{14}C and the final results showed that mechanical homogenization with benzene–isopropanol was most efficient in extracting the residue.

It was claimed by Mumma *et al.*[5] that an exhaustive Soxhlet extraction with methanol–chloroform in an inert atmosphere was necessary to extract a field-incurred residue completely. However, they did not compare this method with the mechanical homogenization procedure using acetonitrile and water as extracting solvents.

Getz[6] showed that acetonitrile could extract both polar and nonpolar organophosphorus residues from vegetable crops. The sample was homogenized with acetonitrile, filtered into a round-bottomed flask and the organic solvent in the filtrate removed by rotary evaporation. The residue and the water left in the flask were transferred to a separating funnel with several small volumes of ethyl acetate. Separation of water and the organic phase was facilitated by adding a small volume of aqueous sodium chloride. The residue was contained in the ethyl acetate layer.

The extract was further purified by chromatography, on an adsorption column consisting of activated charcoal sandwiched between two layers of silica gel, with an eluting solvent of 25% chloroform in ethyl acetate which eluted most of the organophosphate pesticides. Compounds of high aromaticity, such as Guthion, had to be eluted with benzene.

Mills[7] found that acetonitrile could also be used for the direct extraction of organochlorine pesticides from nonfatty foods. The residue was later partitioned into petroleum ether by watering out the acetonitrile with an aqueous sodium chloride solution. The petroleum ether extract was further purified on a Florisil column by eluting with 5, 10 and 15% ethyl ether in petroleum ether.

Extraction of animal tissues was described by de Faubert Maunder *et al.*[8] The sample was minced into small pieces and ground with sand and sodium sulfate until reduced to a flowing powder. It was placed in a beaker and extracted with hexane by boiling on a steam bath. The solvent was decanted and the solid material reextracted in a similar manner with three further aliquots of hexane.

Olson[9] extracted tissue samples with acetonitrile. A finely chopped or ground sample was mechanically homogenized with the solvent and the resulting slurry was filtered. The filtrate was transferred to a separating funnel, while the filter cake and paper were returned to the blender and further homogenized with Skellysolve B, and then refiltered. The filtrates were combined in a separating funnel and partitioned by shaking until two distinct layers were obtained. The bottom layer was usually acetonitrile, with the hydrocarbon on top, although in some cases this arrangement was reversed if the hydrocarbon contained enough dissolved fat to increase significantly its density.

Residues were extracted from matrices such as butter and oils by Eidelman,[10] using dimethyl sulfoxide. The residue contained in the dimethyl sulfoxide was partitioned into petroleum ether by watering out the sulfoxide, and the petroleum ether then chromatographed on Florisil for further isolation of the residue.

Water can be extracted by a continuous method such as that used by Kahn and Wayman.[11] Three continuous liquid–liquid extractors were connected in series, and as the water flowed through the apparatus the residues were continuously extracted into an organic solvent. The flow through the apparatus was maintained at 0.5–1 liter h^{-1}. Suitable solvents would be chloroform, methylene chloride, ethyl acetate or benzene.

A batchwise technique was used by Teasley and Cox,[12] who took large size samples in metal containers and then extracted 1 liter aliquots in a separating funnel with five separate portions of chloroform.

A survey of the different procedures has shown that two more isolation steps are usually necessary after extraction, to ensure that the residue is sufficiently well isolated to be chromatographed.

A first step would be partitioning between a hydrocarbon and a polar solvent that is either immiscible or nearly so in the hydrocarbon. Hydrocarbons that can be used include pentane, petroleum ether, hexane, 2,2,4-trimethylpentane and nonane. Polar solvents that have proved useful in the past are acetonitrile, aqueous methanol, *N*,*N*-dimethylformamide, and dimethyl sulfoxide. In order to keep the lipophilic residue quantitatively in the polar phase, the volume of the polar solvent should be at least twice that of the hydrocarbon. It has been found that three partitionings are usually necessary to transfer the lipophilic residue into the polar phase.

The second step could be liquid–solid adsorption on columns packed with adsorbents such as alumina, silica gel, activated charcoal and Florisil. The sample is placed onto the column with the least polar solvent in use and then a gradient elution, stepwise or with a continuous increase of polar solvent, is used for obtaining the desired pesticides. Appropriate volumes of the respective solvents may be determined experimentally.

Combinations of these various procedures have produced multiresidue techniques, where a number of pesticide residues may be extracted and identified in as few steps as possible.

These, then, are the methods generally used for the monitoring and enforcement of the pesticide residue laws. The earliest chromatographic techniques for assaying residues utilized paper chromatography for separation and identification. Today, thin layer chromatography is rapidly displacing the paper techniques.

3.2 ORGANOCHLORINE INSECTICIDES

Table 3.1 gives the common and chemical names of the organochlorine insecticides mentioned in the text, and the chemical structures of some typical examples are shown in Fig. 3.1.

3.2.1 Paper chromatography

Mitchell[13] in 1958 developed an elaborate paper chromatography system for many organochlorine insecticides. He used Whatman No. 1 chromatographic

TABLE 3.1
Common and chemical names of organochlorine pesticides

Common name	Chemical name
Aldrin	1,2,3,4,10,10-Hexachloro-1,4,4a,5,8,8a-hexahydro-1,4-*endo,exo*-5,8-dimethanonaphthalene
Benzene hexachloride (BHC)	1,2,3,4,5,6-Hexachlorocyclohexane (several isomers)
Captan	*N*-[(Trichloromethyl)thio]-4-cyclohexane-1,2-dicarboximide
Chlorbenside	*p*-Chlorobenzyl-(*p*-chlorophenyl) sulfide
Chlorbenzilate	Ethyl 4,4′-dichlorobenzilate
Chlordane	1,2,4,5,6,7,8,8-Octachloro-3a,4,7,7a-tetrahydro-4,7-methanoindan (60% + 40% related compounds)
CIPC	Isopropyl-3-chlorophenylcarbamate
DDD	2,2-bis(*p*-Chlorophenyl)-1,1-dichloroethane
DDE	1,1-Dichloro-2,2-bis(*p*-chlorophenyl)ethylene
DDT	1,1,1-Trichloro-2,2-bis(*p*-chlorophenyl)ethane
Dichlone (Phygon)	2,3-Dichloro-1,4-naphthaquinone
Dieldrin	1,2,3,4,10,10-Hexachloro-6,7-epoxy-1,4,4a,5,6,7,8,8a-octahydro-1,4-*endo,exo*-5,8-dimethanonaphthalene
Endosulfan	6,7,8,9,10,10-Hexachloro-1,5,5a,6,9,9a-hexahydro-6,9-methano-2,4,3-benzodioxathiepin-3-oxide
Endrin	1,2,3,4,10,10-Hexachloro-6,7-epoxy-1,4,4a,5,6,7,8,8a-octahydro-1,4-*endo,endo*-5,8-dimethanonaphthalene
Heptachlor	1,4,5,6,7,8,8-Heptachloro-3a,4,7,7a-tetrahydro-4,7-methanoindene
Heptachlor epoxide	1,4,5,6,7,8,8-Heptachloro-2,3-epoxy-3a,4,7,7a-tetrahydro-4,7-methanoindan
Kelthane (Dicofol)	1,1-bis(*p*-Chlorophenyl)-2,2,2-trichloroethanol
Kepone	Decachlorooctahydro-1,3,4-metheno-2*H*-cyclobuta[*cd*]pentalen-2-one
Hexachlorobenzene	Hexachlorobenzene
Lindane	1,2,3,4,5,6-Hexachlorocyclohexane (γ isomer)
Methoxychlor	1,1,1-Trichloro-2,2-bis(*p*-methoxyphenyl)ethane
Mirex	Dodecachlorooctahydro-1,3,4-metheno-1*H*-cyclobuta[*cd*]pentalene
Ovex	*p*-Chlorophenyl(*p*-chlorobenzene)sulfonate
PCNB	Pentachloronitrobenzene
Perthane	1,1-Dichloro-2,2-bis(*p*-ethylphenyl)ethane
Perthane olefin	bis(*p*-Ethylphenyl)ethylene
Sulphenone	*p*-Chlorophenyl-phenyl sulfone
Strobane	Terpene polychlorinates of about 66% chlorine content
TCNB (Tecnazene)	1,2,4,5-Tetrachloro-3-nitrobenzene
TDE	1,1-Dichloro-2,2-bis(*p*-chlorophenyl)ethane
TDE olefin	1-Chloro-2,2-bis(*p*-chlorophenyl)ethylene
Tedion (Tetradifon)	*p*-Chlorophenyl-2,4,5-trichlorophenyl sulfone
Telodrin	1,3,4,5,6,7,8,8-Octachloro-1,3,3a,4,7,7a-hexahydro-4,7-methano-isobenzofuran
Tetradifon	*p*-Chlorophenyl-2,4,5-trichlorophenyl sulfone
Thiodan (Endosulfan)	6,7,8,9,10,10-Hexachloro-1,5,5a,6,9,9a-hexahydro-6,9-methano-2,4,3-benzo[*e*]dioxathiepin-3-oxide (α and β)
Toxaphene	Camphene polychlorinates of 67–69% chlorine content

paper in 8″ × 8″ sheets with immobile–mobile phase chromatography. The solvent front was allowed to migrate 15 cm.

Nonaqueous system
Immobile phase: 10 % 2-phenoxyethanol
Mobile phase: 2,2,4-trimethylpentane

R_F: aldrin = 0.18; captan = 0.86; chlordane = 0–0.57; DDT = 0.31; dieldrin = 0.42; endrin = 0.41; heptachlor = 0.32; heptachlor epoxide = 0.30; Kelthane = 0.71; lindane = 0.60; methoxychlor = 0.67; ovex = 0.73, 0.82; Perthane = 0.22; Sulphenone = 0.89; Tedion = 0.49; Thiodan = 0.35, 0.54, 0.89; DDD = 0.48; toxaphene = 0.05–0.67.

Aqueous system
Immobile phase: 5 % refined soybean oil in ethyl ether
Mobile phase: 25 % aqueous 2-methoxyethanol

R_F: aldrin = 0.92; captan = 0.07; chlordane = 0.72–1; *p,p′*-DDT = 0.62; *o,p*-DDT = 0.76; dieldrin = 0.64; endrin = 0.13, 0.63; heptachlor = 0.97; heptachlor epoxide = 0.77; Kelthane = 0–0.07, 0.31; lindane = 0.44; methoxychlor = 0.30; ovex = 0.30; Perthane = 0.93; Sulphenone = 0.12; Tedion = 0.16; Thiodan = 0.36, 0.85; DDD = 0.54; toxaphene = 0.21–0.92.

The compounds were rendered visible by spraying the dried papers after chromatography with a silver nitrate solution (1.7 g of $AgNO_3$ dissolved in the minimum amount of water plus 10 ml of 2-phenoxyethanol and diluted to 200 ml with acetone). The papers were then exposed to strong UV radiation which produced brown to reddish-black spots wherever an organochlorine compound was present.

McKinley and Mahon[14] also utilized immobile–mobile phase chromatography for resolving many organochlorine insecticides. Whatman No. 1 paper was used and R_F values are given relative to lindane = 1.

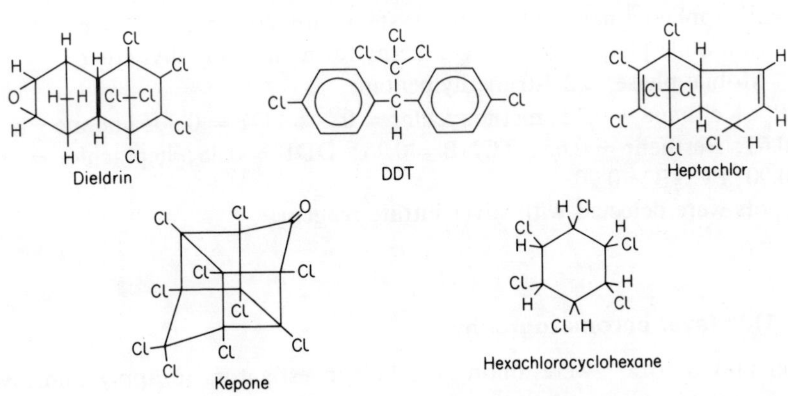

Fig. 3.1. Structural formulae of some organochlorine insecticides showing how diversified their structures can be.

System 1
Immobile: 14% 2-phenoxyethanol in ethyl ether
Mobile: 2,2,4-trimethylpentane

R_F: methoxychlor = 0.67; ovex = 0.46; Perthane = 2.35, 1.93; sulfur = 1.65; Sulphenone = 0.16; aldrin = 2.63; captan = 0.15; chlordane = 1.83; chlorbenside = 1.40; chlorbenzilate = 2.26, 0.52; *o,p'*-DDT = 1.98; *p,p'*-DDT = 1.66; dieldrin = 2.64, 1.55; endrin = 1.48; Kelthane = 0.31, 0.2, 0.11; heptachlor = 2.11.

System 2
Immobile: 4% light mineral oil in ethyl ether
Mobile: 40% aqueous pyridine

R_F: aldrin = 0.10; captan = 1.86; chlordane = 0.18; chlorbenside = 0.32; *o,p'*-DDT = 0.18; *p,p'*-DDT = 0.18; dieldrin = 1.82, 0.33, 0.14; endrin = 0.45; heptachlor = 0.15; Kelthane = 0.73, 0.35, 0.09; methoxychlor = 1.24; ovex = 1; TDE = 0.49; sulfur = 0.0; Sulphenone = 2.36.

System 3
Immobile: 2% light mineral oil in ethyl ether
Mobile: 70% aqueous acetone

R_F: aldrin = 0.27; captan = 0.98; chlorbenside = 0.61; *o,p'*-DDT = 0.55; *p,p'*-DDT = 0.55; dieldrin = 0.08, 0.33, 0.73; endrin = 0.71; heptachlor = 0.38; methoxychlor = 1.09; TDE = 0.85; sulfur = 1.10; Sulphenone = 2.42.

Two reagents were used for making the migrated spots visible: the silver nitrate solution described by Mitchell,[13] and a 0.002M solution of *N*-bromosuccinimide in methyl chloroform which was followed by a second treatment with an alkaline solution and a 0.003M ethanolic solution of fluorescein. The spots were visible as white areas on a pink background, which when observed in black light (filtered UV) appeared as dark blue quenched spots on a yellow fluorescent background.

Mills[15] applied a new nonaqueous system with Whatman No. 1 paper.

Immobile phase: 35% *N,N*-dimethylformamide in ethyl ether
Mobile phase: 2,2,4-trimethylpentane

R_F: Kelthane = 0.22; methoxychlor = 0.26; TDE = 0.30; lindane = 0.63; Perthane = 0.65; TCNB = 0.75; DDE = 0.85; heptachlor = 0.90; PCNB = 0.90.

The spots were detected with silver nitrate reagent.

3.2.2 Thin layer chromatography

Walker and Beroza[16] were among the first investigators to apply thin layer chromatography to the analysis of pesticides. They coated glass plates with Silica Gel G and used several solvent systems to run the chromatograms.

System 1: hexane
R_F: aldrin = 0.43; chlordane = 0.25–0.38; *p,p'*-DDE = 0.28; *o,p'*-DDT = 0.22; *p,p'*-DDT = 0.19; dieldrin = 0.19; heptachlor = 0.35; heptachlor epoxide = 0.07; Kelthane = 0.03; Kepone = 0.0; lindan = 0.09; methoxychlor = 0.02; mirex = 0.51.

System 2: hexane–acetone (9:1).
R_F: aldrin = 0.67; captan = 0.17; chlordane = 0.57–0.65; *p,p'*-DDE = 0.62; *o, p'*-DDT = 0.57; *p,p'*-DDT = 0.54; dieldrin = 0.49; endrin = 0.54; heptachlor = 0.65; heptachlor epoxide = 0.55; Kelthane = 0.34; lindane = 0.42; methoxychlor = 0.32; mirex = 0.73.

System 3: hexane–acetone (8:2)
R_F: aldrin = 0.76; captan = 0.37; *p,p'*-DDE = 0.64; *o,p'*-DDT = 0.58; *p,p'*-DDT = 0.58; Kelthane = 0.51; Kepone = 0.17; methoxychlor = 0.45.

System 4: hexane–ethyl acetate (9:1)
R_F: aldrin = 0.62; captan = 0.05; chlordane = 0.52–0.64; *p,p'*-DDE = 0.48; *o,p'*-DDT = 0.38; *p,p'*-DDT = 0.38; dieldrin = 0.37; endrin = 0.35; heptachlor = 0.56; heptachlor epoxide = 0.40; Kelthane = 0.38; lindane = 0.37; methoxychlor = 0.29; mirex = 0.68.

System 5: benzene
R_F: aldrin = 0.80; captan = 0.15; *p,p'*-DDE = 0.68; *o,p'*-DDT = 0.66; *p,p'*-DDT = 0.68; dieldrin = 0.55; heptachlor = 0.74; heptachlor epoxide = 0.65; Kelthane = 0.59; lindane = 0.66; methoxychlor = 0.50; mirex = 0.74; TDE = 0.76.

Residues were rendered visible with the silver nitrate reagent used by Mitchell.[13]

Brandt[17] coated plates with a TLC grade of porous glass and used a number of solvent systems to resolve the following organochlorine insecticides.

Solvent system A: chloroform
R_F: aldrin = 0.82; *p,p'*-DDT = 0.82; lindane = 0.70; endrin = 0.60; heptachlor = 0.87; dieldrin = 0.65; methoxychlor = 0.57.

Solvent system B: petroleum ether–chloroform (90:10)
R_F: aldrin = 0.60; *p,p'*-DDT = 0.40; lindane = 0.12.

Solvent system C: benzene
R_F: aldrin = 0.85; *p,p'*-DDT = 0.85; lindane = 0.67; endrin = 0.20; heptachlor = 0.85; dieldrin = 0.45; methoxychlor = 0.37.

Solvent system D: pentane–petroleum ether (10:90)
R_F: aldrin = 0.40; *p,p'*-DDT = 0.20; lindane = 0.05; heptachlor = 0.30.

Solvent system E: pentane–benzene (90:10)
R_F: aldrin = 0.90; *p,p'*-DDT = 0.45; lindane = 0.15; endrin = 0.10; heptachlor = 0.60; dieldrin = 0.10.

Solvent system F: pentane–benzene (10:90)

R_F: aldrin = 0.87; *p,p'*-DDT = 0.87; lindane = 0.67; endrin = 0.45; heptachlor = 0.87; dieldrin = 0.45; methoxychlor = 0.35.

Solvent system G: pentane–benzene (50:50)

R_F: aldrin = 0.87; *p,p'*-DDT = 0.78; lindane = 0.50; endrin = 0.30; heptachlor = 0.87; dieldrin = 0.30; methoxychlor = 0.20.

Solvent system H: petroleum ether–acetone (95:5)

R_F: aldrin = 1; *p,p'*-DDT = 0.50; lindane = 0.2; endrin = 0.27; heptachlor = 0.7; dieldrin = 0.25; methoxychlor = 0.10.

Solvent system I: petroleum ether–acetone (90:10)

R_F: aldrin = 0.95; *p,p'*-DDT = 0.70; lindane = 0.45; endrin = 0.65; heptachlor = 0.9; dieldrin = 0.6; methoxychlor = 0.35.

The spots were detected by spraying the plates with a solution of *N,N'*-dimethylamino-*p*-phenylenediamine hydrochloride (0.5 g dissolved in 100 ml of sodium methoxide solution), and then placing them in a tank containing iodine vapor. The resulting spots were brownish-black in color.

Kovacs[18] used microslides 3″ × 2″ coated with Aluminum Oxide G for resolving organochlorine insecticides.

Solvent: *n*-heptane

R_F: hexachlorobenzene = 0.89; aldrin = 0.86; DDE = 0.80; heptachlor = 0.78; TCNB = 0.70; Telodrin = 0.69; DDT = 0.63; BHC = 0.50, 0.41, 0.11, 0.03; chlorbenside = 0.48; Perthane = 0.47; lindane = 0.40; DDD = 0.36; Thiodan = 0.28; endrin = 0.25; heptachlor epoxide = 0.23; dieldrin = 0.18; methoxychlor = 0.12; ovex = 0.05; CIPC = 0.01; Kelthane = 0.01.

Immobile phase: 25% *N,N*-dimethylformamide in ethyl ether

Mobile phase: 2,2,4-trimethylpentane

R_F: hexachlorobenzene = 0.90; aldrin = 0.81; DDE = 0.50; heptachlor = 0.60; TCNB = 0.45; Telodrin = 0.49; DDT = 0.25; BHC = 0.21, 0.16, 0.11, 0.05; chlorbenside = 0.23; Perthane = 0.32; lindane = 0.15; DDD = 0.11; Thiodan = 0.45, 0.14; endrin = 0.42; heptachlor epoxide = 0.30; dieldrin = 0.37; methoxychlor = 0.10; ovex = 0.08; CIPC = 0.09; Kelthane = 0.22.

Silver nitrate reagent was used for detecting the spots.

Thomas *et al.*[19] used aluminium oxide layers in conjunction with various solvents and used *p,p'*-TDE as the reference compound.

Solvent system 1: *n*-heptane

p,p'-TDE migrated a distance of 3.9 cm. R_{TDE}: hexachlorobenzene = 2.7; aldrin = 2.1; *p,p'*-DDE = 2.0; chlordane = 2.01, 1.8, 1.4, 1.2; Perthane olefin = 1.8; *p,p'*-TDE olefin = 1.8; Telodrin = 1.7; toxaphene = 1.7, 1.2; *p,p'*-DDT = 1.6; *o,p'*-TDE olefin = 1.6; chlorbenside = 1.3; γ BHC = 1.3; Perthane = 1.3; lindane = 1.1; *o,p'*-TDE = 1.1; endosulfan (A and B) = 0.88, 0.00; heptachlor epoxide = 0.71; endrin = 0.71; dieldrin = 0.52; methoxychlor = 0.33, 0.27;

β-BHC = 0.27; ovex = 0.18; dichlone = 0.16; tetradifon = 0.11; θ-BHC = 0.10; Kelthane = 0.06.

Solvent system 2: heptane–acetone (98:2)

p,p′-TDE migrated a distance of 5.7 cm. R_{TDE}: hexachlorobenzene = 1.7; aldrin = 1.4; *p,p′*-DDE = 1.4; heptachlor = 1.4; chlordane = 1.4, 1.3, 1.2, 1.1; Telodrin = 1.4; toxaphene = 1.3, 1.2; strobane = 1.3, 1.2; *p,p′*-DDT = 1.2; chlorbenside = 1.2; α-BHC = 1.1; Perthane = 1.2; lindane = 0.02; *o,p′*-TDE = 0.95; endosulfan (A and B) = 0.92, 0.24; heptachlor epoxide = 1.0; endrin = 1.0; dieldrin = 0.90; methoxychlor = 0.79; β-BHC = 0.72; ovex = 0.76; dichlone = 0.72; tetradifon = 0.82; θ-BHC = 0.25; Kelthane = 0.28; sulphenone = 0.31.

The compounds were made visible with silver nitrate reagent and UV irradiation.

3.3 ORGANOPHOSPHORUS PESTICIDES

This group of compounds was a byproduct of chemical research into toxic nerve and blood gases during World War II. They are essentially di- and tri-esters of phosphonic, phosphoric, phosphorothioic, and phosphorodithioic acids. Their primary effect as insecticides is the inhibition of the cholinesterase enzymes and the consequent short circuiting of the nervous system caused by the accumulation of an excess of unhydrolyzed acetylcholine at nerve endings. Some of these pesticides are extremely toxic both to mammals and to the lower-animal pests to be controlled, while others are relatively harmless to mammals but very effective against the pests.

Figure 3.2 shows the basic structures of the pesticides with some examples of specific compounds. Table 3.2 contains the common and chemical names of the organophosphorus pesticides used in the text.

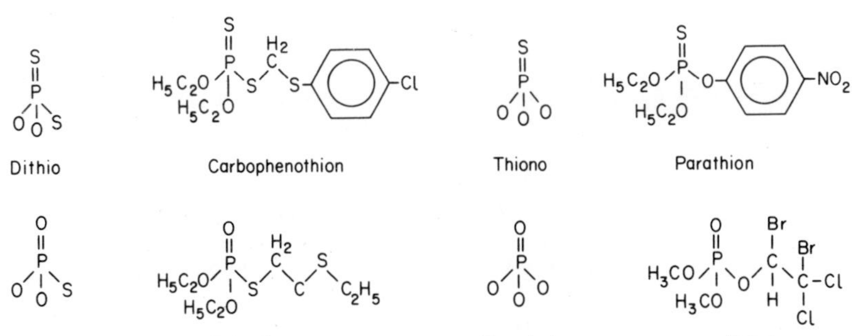

Fig. 3.2. Structural formulae of the different types of organophosphate pesticides, with some specific examples.

TABLE 3.2
Common and chemical names of organophosphate pesticides

Common name	Chemical name
Azinphos-ethyl	*O,O*-Diethyl-*S*-[(4-oxo-1,2,3-benzotriazin-3(4*H*)-yl)methyl]-phosphorodithioate
Azinphos-methyl	*O,O*-Dimethyl-*S*-[(4-oxo-1,2,3-benzotriazin-3(4*H*)-yl)methyl]-phosphorodithioate
Bensulide	*O,O*-Diisopropylphosphorodithioate *S*-ester with *N*-(2-mercapto-ethyl)benzenesulfonamide
Bidrin	See dicrotophos
Bomyl	Dimethyl-1,3-di-carboxymethoxy-1-propen-2-yl phosphate
Bromophos	*O*-(4-Bromo-2,5-dichlorophenyl)-*O,O*-dimethyl phosphorothioate
Carbophenothion	*S*-[((*p*-Chlorophenyl)thio)methyl]-*O,O*-diethyl phosphorodithioate
Chlorpyrifos	*O,O*-Diethyl-*O*-(3,5,6-trichloro-2-pyridyl)phosphorothioate
Ciodrin	See crotoxyphos
Chlorthion	*O,O*-Dimethyl-*O*-(3-chloro-4-nitrophenyl)phosphorothioate
Coumaphos	*O,O*-Diethyl-*O*-(3-chloro-4-methyl-2-oxo-2*H*-1-benzopyran-7-yl)-phosphorothioate
Coumaphos oxygen analog	*O,O*-Diethyl-*O*-(3-chloro-4-methyl-2-oxo-2*H*-1-benzopyran-7-yl)-phosphate
Crotoxyphos	α-Methylbenzyl-(*E*)-3-hydroxycrotonate dimethyl phosphate
Crufomate	4-*tert*-Butyl-2-chlorophenylmethyl-methylphosphoroamidate
Co-Ral	See coumaphos
Dasanit	See fensulfothion
DEF	*S,S,S*-Tributylphosphorotrithioate
Delnav	See dioxathion
Demeton	*O,O*-Diethyl-*O*(and *S*)-[2-(ethylthio)ethyl]phosphorothioates
Demeton-*O*-methyl	See thiol isomer of demeton
Demeton-*S*-methyl	See thiono isomer of demeton
Demeton thiol	The *O*-analog of demeton
Demeton thiol sulfone	*O,O*-Diethyl-*O*-(2-ethylsulfonylethyl)phosphorothioate
Diazinon	*O,O*-Diethyl-*O*-(2-isopropyl-6-methyl-4-pyrimidinyl)phosphoro-thioate
Dicapthon	*O*-(2-Chloro-4-nitrophenyl)-*O,O*-dimethylphosphorothioate
Dichlofenthion	*O*-(2,4-Dichlorophenyl)-*O,O*-diethyl phosphorothioate
Dichlorvos	2,2-Dichlorovinyl dimethylphosphate
Dicrotophos	Dimethylphosphate ester with (*E*)-3-hydroxy-*N,N*-dimethyl-crotonamide
Dimethoate	*O,O*-Dimethyl-*S*-(methylcarbamoylmethyl)phosphorodithioate
Dimethoate oxygen analog	*O,O*-Dimethyl-*O*-*S*-(methylcarbamoylmethyl)phosphorothioate
Dioxathion	*S,S′*-*p*-Dioxane-2,3-diyl-*O,O,O′,O′*-tetraethyl bis(phosphoro-dithioate)
Dipterex	See trichlorfon
Disulfoton	*O,O*-Diethyl-*S*-[2-(ethylthio)ethyl]phosphorodithioate
Disulfoton sulfoxide	*O,O*-Diethyl-*S*-(2-ethylsulfinylethyl)phosphorodithioate
Disulfoton sulfone	*O,O*-Diethyl-*S*-(2-ethylsulfonylethyl)phosphorodithioate
Di-Syston	See disulfoton
EPN	*O*-Ethyl-*O*-(*p*-nitrophenyl)phenylphosphorothioate
Ethion	*O,O,O′,O′*-Tetraethyl-*S,S′*-methylene-bis(phosphorodithioate)
Famphur	*O*-[*p*-(Dimethylsulfamoyl)phenyl]-*O,O*-dimethylphosphorothioate
Fenchlorphos	*O,O*-Dimethyl-*O*-(2,4,5-trichlorophenyl)phosphorothioate
Fenitrothion	*O,O*-Dimethyl-*O*-(4-nitro-*m*-tolyl)phosphorothioate
Fensulfothion	*O,O*-Diethyl-*O*-[*p*-(methylsulfinyl)phenyl]phosphorothioate

Common name	Chemical name
Fenthon	*O,O*-Dimethyl-*O*-[4-(methylthio)-*m*-tolyl]phosphorothioate
Fonofos	*O*-Ethyl-*S*-phenyl ethylphosphorodithioate
Formothion	*O,O*-Dimethylphosphorodithioate-*S*-ester with *N*-formyl-2-mercapto-*N*-methylacetamide
Guthion	See azinphos methyl
Imidan	See phosmet
Leptophos	*O*-(4-Bromo-2,5-dichlorophenyl)-*O*-methylphenylphosphorothioate
Malaoxon	*O,O*-Dimethyl-*S*-[1,2-di-(ethoxycarbonyl)ethyl]phosphorothioate
Malathion	Diethylmercaptosuccinate-*S*-ester with *O,O*-dimethylphosphorodithioate
Menazon	*O,O*-Dimethyl-*S*-(4,6-diamino-1,3,5-triazin-2-ylmethyl)phosphorodithioate
Methyl parathion	*O,O*-Dimethyl-*O*-(*p*-nitrophenyl)phosphorothioate
Me-Trithion	See carbophenothion
Mevinphos	Methyl-(*E*)-3-hydroxycrotonate dimethylphosphate
Monocrotophos	Dimethylphosphate ester with (*E*)-3-hydroxy-*N*-methylcrotonamide
Naled	1,2-Dibromo-2,2-dichloroethyl dimethylphosphate
Nemacide	See dichlofenthion
Oxydemeton-methyl	*S*-[2-(Ethylsulfinyl)ethyl]-*O,O*-dimethylphosphorothioate
Paraoxon	*O,O*-Diethyl-*O*-(*p*-nitrophenyl)phosphate
Parathion	*O,O*-Diethyl-*O*-(*p*-nitrophenyl)phosphorothioate
Phenkapton	*O,O*-Diethyl-*S*-(2,5-dichlorophenylthiomethyl)phosphorodithioate
Phorate	*O,O*-Diethyl-*S*-[ethylthio)methyl]phosphorodithioate
Phorate sulfoxide	*O,O*-Diethyl-*S*-(ethylsulfinylmethyl)phosphorodithioate
Phorate sulfone	*O,O*-Diethyl-*S*-(ethylsulfonylmethyl)phosphorodithioate
Phoratoxon	*O,O*-Diethyl-*O*-(ethylthiomethyl)phosphorothioate
Phoratoxon sulfoxide	*O,O*-Diethyl-*O*-(ethylsulfinylmethyl)phosphorothioate
Phoratoxon sulfone	*O,O*-Diethyl-*O*-(ethylsulfonylmethyl)phosphorothioate
Phosdrin	See mevinphos
Phosmet	*O,O*-Dimethylphosphorodithioate-*S*-ester with *N*-(mercaptomethyl)phthalimide
Phosphamidon	2-Chloro-2-(diethylcarbamoyl)-1-methylvinyl dimethylphosphate
Ronnel	*O,O*-Dimethyl-*O*-(2,4,5-trichlorophenyl)phosphorothioate
Rogor	See dimethoate
Ruelene	See crufomate
Sulfotepp	*O,O,O′,O′*-Tetraethyl dithiopyrophosphate
TEPP	Tetraethylpyrophosphate
Thimet	See phorate
Thionazin	*O,O*-Diethyl-*O*-pyrazinyl phosphorothioate
Trichlorfon	Dimethyl-(2,2,2-trichloro-1-hydroxyethyl)phosphonate
Trithion	See carbophenothion
Vamidothion	*O,O*-Dimethyl-*S*-[2-(1-methylcarbamoylethylthio)ethyl]phosphorothioate
Vamidothion sulfoxide	*O,O*-Dimethyl-*S*-[2-(1-methylcarbamoylethylsulfinyl)ethyl]-phosphorothioate
VC-13	See dichlofenthion
Zinophos	See thionazin
Zytron	*O*-(2,4-Dichlorophenyl)-*O′*-methyl-*N*-isopropylphosphoramidothioate

3.3.1 Paper chromatography

Mitchell[20] applied immobile–mobile phase paper chromatography for the resolution of some organophosphorus insecticides.

Adsorbent: Whatman No. 1 8″ × 8″ paper
Immobile phase: 20% *N,N*-dimethylformamide in ethyl ether
Mobile phase: 2,2,4-trimethylpentane

R_F: Delnav (*cis* and *trans*) = 0.80, 0.95; diazinon = 0.48, 0.97; EPN = 0.69; Guthion = 0.10; malathion = 0.62; methyl parathion = 0.35; parathion = 0.75; phosdrin (α and β) = 0.11, 0.04; ronnel = 0.96; demeton = 0.96, 0.59 (thiono and thiol isomers, respectively); Trithion = 0.96.

Immobile phase: 10% mineral oil in ethyl ether
Mobile phase: 50% *N,N*-dimethylformamide in water

R_F: Delnav (*cis* and *trans*) = 0.18, 0.39; diazinon = 0.24, 0.96; EPN = 0.34; Guthion = 0.89; malathion = 0.81; methyl parathion = 0.81; parathion = 0.51; phosdrin = 0.98; ronnel = 0.12; demeton = 0.34, 0.90 (thiono and thiol isomers, respectively); Trithion = 0.06.

The dry chromatogram was placed in a tank containing bromine fumes for several seconds, then removed and sprayed with a 0.025% solution of fluorescein in ethanol. Bromine converts fluorescein to eosin which fluoresces yellow. Where a residue spot absorbed the bromine, there was no conversion to eosin and the migrated spots were visible as dark blue quenched areas on a fluorescent yellow background.

MacRae and McKinley[21] used acetylated paper in conjunction with immobile–mobile phase chromatography.

Adsorbent: Schleicher and Schuell acetylated paper
Immobile phase: 2% mineral oil in ethyl ether
Mobile phase: 70% aqueous acetone

R_F: Co-Ral = 0.23; diazinon = 0.38; Dipterex = 0.25; Di-Syston = 0.23; EPN = 0.25; ethion = 0.22; Guthion = 0.30; malathion = 0.34; Nemacide = 0.22; parathion = 0.24; phosdrin = 0.56; dimethoate = 0.54; demeton = 0.32; 0.46 (thiono and thiol isomers, respectively); TEPP = 0.71; Thimet = 0.25; Trithion = 0.20.

The spots were made visible with silver nitrate reagent and the bromine–fluorescein treatment described earlier.

Bates[22] found that Whatman No. 20 chromatographic paper gave smaller, more definite spots and therefore greater resolution.

Immobile phase: 30% *N,N*-dimethylformamide in acetone
Mobile phase: hexane

R_F: Phenkapton = 0.98; Trithion = 0.98; ethion = 0.98; diazinon = 0.98; phorate = 0.96; fenthion = 0.91; malathion = 0.85; parathion = 0.82; azinphos-ethyl = 0.63; azinphos-methyl = 0.92; mevinphos = 0.92; formothion = 0.96; menazon = 0.78;

oxydemeton-methyl = 0.90; vamidothion = 0.95; dimethoate = 0.95; vamidothion sulfoxide = 0.97.

The spots were made visible by spraying with a 0.5 solution of 2,6-dibromo-*N*-chloro-*p*-quinoneimine in cyclohexane. Subsequent heating made the residue sites turn orange or yellow.

3.3.2 Thin layer chromatography

Walker and Beroza[16] separated organophosphorus pesticides on Silica Gel G layers.

Solvent system 1: hexane–diethyl ether (9:1)

R_F: carbophenothion = 0.43; Chlorthion = 0.13; Co-Ral = 0.06; Delnav = 0.07; demeton (thiono isomer) = 0.34; Dicapthon = 0.18; Di-Syston = 0.41; Di-Syston sulfoxide = 0.0; EPN = 0.24; ethion = 0.34; malathion = 0.04; methyl parathion = 0.13; parathion = 0.23; phorate = 0.41; sulfotepp = 0.31; VC-13 = 0.46; Zectran = 0.01.

Solvent system 2: hexane–acetone (8:2)

R_F: carbophenothion = 0.60; Chlorthion = 0.15; Co-Ral = 0.23; Delnav = 0.43; demeton (thiono isomer) = 0.61; demeton (thiol isomer) = 0.31; Dicapthon = 0.38; Di-Syston = 0.61; Di-Syston sulfone = 0.23; EPN = 0.48; ethion = 0.57; Guthion = 0.27; malathion = 0.35; methyl parathion = 0.34; parathion = 0.44; phorate = 0.60; sulfotepp = 0.42; VC-13 = 0.52; Zectran = 0.38.

Solvent system 3: hexane–methanol (9:1)

R_F: carbophenothion = 0.52; Chlorthion = 0.15; Co-Ral = 0.19; Delnav = 0.18; demeton (thiono isomer) = 0.52; demeton (thiol isomer) = 0.14; Dicapthon = 0.24; Di-Syston = 0.54; EPN = 0.41; ethion = 0.50; Guthion = 0.13; malathion = 0.25; methyl parathion = 0.29; parathion = 0.38; phorate = 0.53; sulfotepp = 0.44; VC-13 = 0.55; Zectran = 0.10.

Solvent system 4: benzene

R_F: carbophenothion = 0.65; Chlorthion = 0.53; Co-Ral = 0.22; Delnav = 0.27; demeton (thiono isomer) = 0.43; Dicapthon = 0.53; Di-Syston = 0.52; EPN = 0.53; ethion = 0.56; Guthion = 0.06; malathion = 0.08; methyl parathion = 0.41; parathion = 0.47; phorate = 0.54; sulfotepp = 0.47; VC-13 = 0.59; Zectran = 0.07.

Solvent system 5: benzene–diethyl ether (9:1)

R_F: carbophenothion = 0.77; Chlorthion = 0.68; Co-Ral = 0.60; Delnav = 0.67; demeton (thiono isomer) = 0.72; demeton (thiol isomer) = 0.08; diazinon = 0.08; Dicapthon = 0.67; Di-Syston = 0.73; EPN = 0.70; ethion = 0.74; Guthion = 0.36; malathion = 0.51; methyl parathion = 0.67; parathion = 0.72; phorate = 0.76; sulfotepp = 0.72; VC-13 = 0.74; Zectran = 0.27.

Solvent system 6: benzene–acetone (9:1)

R_F: carbophenothion = 0.76; Chlorthion = 0.67; Co-Ral = 0.61;

Delnav = 0.73; demeton (thiono isomer) = 0.73; demeton (thiol isomer) = 0.30; diazinon = 0.33; Dicapthon = 0.70; Di-Syston = 0.76; EPN = 0.73; ethion = 0.77; Guthion = 0.54; malathion = 0.63; methyl parathion = 0.67; parathion = 0.71; phorate = 0.75; sulfotepp = 0.66; VC-13 = 0.68; Zectran = 0.44.

Solvent system 7: benzene-methanol (9:1)

R_F: carbophenothion = 0.81; Chlorthion = 0.75; Co-Ral = 0.67; Delnav = 0.82; demeton (thiono isomer) = 0.81; demeton (thiol isomer) = 0.54; diazinon = 0.54; Dicapthon = 0.78; dimethoate = 0.28; Di-Syston = 0.80; EPN = 0.79; ethion = 0.80; Guthion = 0.69; malathion = 0.74; methyl parathion = 0.75; parathion = 0.76; phorate = 0.78; sulfotepp = 0.70; VC-13 = 0.73; Zectran = 0.48.

The spots were made visible with two alternative sets of chromogenic reagents; bromine with fluorescein, and silver nitrate solution with exposure to UV radiation. The bromine was applied as a 5% solution in carbon tetrachloride in this instance instead of the vapor form.

Kovacs[23] applied immobile–mobile phase chromatography in conjunction with an inorganic adsorbent.

Adsorbent: Aluminum Oxide G

Immobile phase: 15% *N,N*-dimethylformamide in ethyl ether

Mobile phase: methylcyclohexane

R_F: Rogor = 0.01; Guthion = 0.09; methyl parathion = 0.17; Co-Ral = 0.23; malathion = 0.34; Delnav = 0.37; parathion = 0.41; Systox (thiol) = 0.44; EPN = 0.49; Methyl Trithion = 0.50; sulfotepp = 0.69; Trithion = 0.74; ronnel = 0.76; ethion = 0.77; Systox (thiono) = 0.79; Thimet = 0.81; Di-Syston = 0.82; diazinon = 0.86.

The spots were made visible by spraying the dry plate with a 0.2% solution of tetrabromophenolphthalein ethyl ester in acetone. The plate was then oversprayed with a 0.5% aqueous acetone solution of silver nitrate, heated in a forced-draft oven for 5–10 min at 105 °C, and oversprayed again with a 5% acetone solution of citric acid. Blue or purple spots of the pesticide appeared against a yellow background.

Getz and Wheeler[24] used silica gel and Florisil to resolve a number of organophosphorus pesticides.

Adsorbent: SilicAr 4

System 1: 2,2,4-trimethylpentane–acetone–chloroform (70:25:5)

R_F: carbophenothion = 0.64; dioxathion = 0.18; dimethoate = 0.0; azinphos-ethyl = 0.08; azinphos-methyl = 0.06; Imidan = 0.06; malathion = 0.10; ethion = 0.41; Methyl Trithion = 0.44; phorate = 0.54; disulfoton = 0.51; Dasanit = 0.07; fenthion = 0.69; Dicapthon = 0.59; EPN = 0.45; methyl parathion = 0.56; parathion = 0.75; ronnel = 0.59; Zytron = 0.28; Zinophos = 0.37;

diazinon = 0.29; coumaphos = 0.22; demeton = 0.68; malaoxon = 0.20; demeton thiol sulfone = 0.10.

Solvent system 2: cyclohexane–acetone–chloroform (70:25:5)

R_F: carbophenothion = 0.96; dioxathion = 0.88; dimethoate = 0.19; azinphos-ethyl = 0.68; azinphos-methyl = 0.57; Imidan = 0.57; malathion = 0.74; ethion = 0.92; Methyl Trithion = 0.90; phorate = 1; disulfoton = 1; Dasanit = 0.33; fenthion = 0.90; Dicapthon = 0.85; EPN = 0.90; methyl parathion = 0.77; parathion = 0.81; ronnel = 0.93; Zytron = 0.79; Zinophos = 0.92; diazinon = 0.82; coumaphos = 0.72; demeton = 0.94; demeton thiol sulfone = 0.10; malaoxon = 0.20; coumaphos oxygen analog = 0.26; trichlorfon = 0.09; Ruelene = 0.23; Bomyl = 0.13; Ciodrin = 0.31; dichlorvos = 0.42; naled = 0.53; mevinphos = 0.23; phosphamidon = 0.13; Bidrin = 0.03.

Solvent system 3: 2,2,4-trimethylpentane–acetonitrile–ethyl acetate (20:20:15)

R_F: trichlorfon = 0.26; Ruelene = 0.45; Bomyl = 0.35; Bidrin = 0.09; Ciodrin = 0.41; dichlorvos = 0.59; naled = 0.56; mevinphos = 0.39; phosphamidon = 0.25.

Solvent system 4: cyclohexane–isopropyl ether–acetone (20:40:40)

R_F: dimethoate = 0.64; Dasanit = 1.0; dimethoate oxygen analog = 0.06; oxydemeton-methyl = 0.06; demeton thiol sulfone = 0.64; malaoxon = 0.87; coumaphos oxygen analog = 0.93; trichlorfon = 0.72; Ruelene = 0.91; Bomyl = 0.83; Bidrin = 0.22; Ciodrin = 1.0; dichlorvos = 1.0; naled = 1.0; mevinphos = 0.86; phosphamidon = 0.67.

When SilicAr 7 was used as the adsorbent significantly different R_F values were obtained with solvent systems 1 and 3.

Solvent 1: R_F: carbophenothion = 0.89; dioxathion = 0.31; dimethoate = 0.02; azinphos-ethyl = 0.17; azinphos-methyl = 0.14; Imidan = 0.17; malathion = 0.27; ethion = 0.72; Methyl Trithion = 0.82; phorate = 0.93; disulfoton = 0.87; Dasanit = 0.0; fenthion = 0.43; Dicapthon = 0.46; EPN = 0.34; methyl parathion = 0.45; parathion = 0.68; ronnel = 0.50; Zytron = 0.18; Zinophos = 0.26; diazinon = 0.22; coumaphos = 0.16; demeton = 0.57; malaoxon = 0.34; demeton thiol sulfone = 0.0.

Solvent 3. R_F: coumaphos oxygen analog = 0.37; trichlorfon = 0.25; Ruelene = 0.39; Bomyl = 0.29; Bidrin = 0.09; Ciodrin = 0.34; dichlorvos = 0.37; naled = 0.41; mevinphos = 0.29; phosphamidon = 0.20.

Adsorbent: Florisil

Solvent system 1: 2,2,4-trimethylpentane–acetone–chloroform (70:25:5)

R_F: carbophenothion = 0.90; dioxathion = 0.30; dimethoate = 0.03; azinphos-ethyl = 0.12; azinphos-methyl = 0.09; Imidan = 0.14; malathion = 0.24; ethion = 0.72; Methyl Trithion = 0.72; phorate = 0.86; disulfoton = 0.86; Dasanit = 0.0; fenthion = 0.64; Dicapthon = 0.45; EPN = 0.49; methyl parathion = 0.33; parathion = 0.48; ronnel = 0.77; Zytron = 0.68; Zinophos = 0.27; diazinon = 0.47; coumaphos = 0.12; demeton = 0.20; dimethoate oxygen analog = 0.52.

Solvent system 2: cyclohexane–acetone–chloroform (70:25:5)
R_F: carbophenothion = 0.93; dioxathion = 0.81; dimethoate = 0.30; azinphos-ethyl = 0.69; azinphos-methyl = 0.60; Imidan = 0.60; malathion = 0.75; ethion = 0.85; Methyl Trithion = 0.85; phorate = 0.88; disulfoton = 0.88; Dasanit = 0.38; fenthion = 0.89; Dicapthon = 0.83; EPN = 0.87; methyl parathion = 0.77; parathion = 0.86; ronnel = 0.92; Zytron = 0.91; Zinophos = 0.78; diazinon = 0.88; coumaphos = 0.71; demeton = 0.77; dimethoate oxygen analog = 0.07; demeton thiol sulfone = 0.24; malaoxon = 0.44; coumaphos oxygen analog = 0.32; trichlorfon = 0.17; Ruelene = 0.30; Bidrin = 0.13; Ciodrin = 0.44; mevinphos = 0.38; phosphamidon = 0.22.

Solvent system 3: acetone–ispropyl ether–cyclohexane (40:40:20)
R_F: dimethoate = 0.82; Dasanit = 0.65; dimethoate oxygen analog = 0.23; oxydemeton-methyl = 0.11; malaoxon = 0.19; trichlorfon = 0.62; Ruelene = 0.89; Bidrin = 0.44; phosphamidon = 0.74.

Three chromogenic reagents were used:

(i) Silver nitrate–bromcresol green for detecting the thiophosphoryl groups. Five ml of a 1% methanolic solution of bromcresol green was added to 95 ml of a 2% methanolic solution of silver nitrate. The plate was sprayed with the dye solution and then dried in a forced-draft oven at 100 °C. When cool it was sprayed over with a pH 4.4 citrate or acetate buffer solution, redried in the oven and observed. The background turned yellow and a bluish-black or magenta spot indicated the presence of an organophosphate pesticide that had a thiophosphoryl configuration.

(ii) The plate was sprayed with a 2% acetone solution of 4-(*p*-nitrobenzyl)pyridine and then heated at 110 °C for 10 min. It was then oversprayed with a 10% acetone solution of tetraethylenepentamine. The pesticides appeared as purple-blue spots on a white background. This test detected all types of organophosphates.

(iii) A 8″ × 8″ sheet of Whatman No. 3MM paper was sprayed with a solution of 20 ml of horse serum (Difco) + 15 ml of indicator (1.2% bromthymol blue in 0.1N sodium hydroxide) + 2 ml of 0.1N sodium hydroxide + 60 ml of water. This was done until the excess spray ran off. The sheet was then placed in contact with the thin layer plate, another glass plate placed on top and weighted down, and 30 min contact time allowed for reaction. The paper was then separated from the plate and sprayed with a 2% solution of acetylcholine chloride. The residue spot was visible as blue on a yellow background. This reagent detected all phosphates with enzyme-inhibiting properties.

Abbott *et al.*[25] showed that small silica gel plates of 15 × 7.5 cm could be used for screening purposes and the pesticide quantified by removing it from the plate.

The plates were developed with hexane and acetone (5:1), and the following R_F values were obtained: azinphos-methyl = 0.18; carbophenothion = 0.65; Chlorthion = 0.34; demeton-*O*-methyl = 0.50; demeton-*S*-methyl = 0.13; diazinon = 0.49; dimethoate = 0.03; disulfoton = 0.65; Fenchlorphos = 0.60; malathion = 0.33; paraoxon = 0.11; parathion = 0.45; phenkapton = 0.62; phorate = 0.65; phosphamidon = 0.37.

The compounds were removed from the silica gel by scraping off the zone containing the spot, and then quantified by a modified phosphomolybdenum blue procedure.[26]

Blinn[27] applied thin layer chromatography to the identification of the metabolites of the pesticide Phorate.

Adsorbent: Silica Gel G

Solvent system A: methanol–chloroform (1.75:98.25)

R_F: phorate = 0.87; phorate sulfoxide = 0.37; phorate sulfone = 0.69; phoratoxon = 0.55; phoratoxon sulfoxide = 0.10; phoratoxon sulfone = 0.47.

Solvent system B: nitromethane–acetonitrile–toluene (15:40:45)

R_F: phorate = 0.88; phorate sulfoxide = 0.45; phorate sulfone = 0.74; phoratoxon = 0.60; phoratoxon sulfoxide = 0.11; phoratoxon sulfone = 0.47.

The residues were detected by reacting with a solution of 5 ml of 5% palladium chloride and 1 ml of concentrated hydrochloric acid diluted to 100 ml in ethanol. Yellow to brown spots were formed, which sometimes took several hours to become visible. The resulting spots could be scraped off, eluted and an IR spectrum recorded for the metabolite (the complexed form did not interfere with the IR spectrum of the residue).

3.4 CARBAMATES

Carbamate pesticides also inhibit cholinesterase, but they are less toxic to mammals than the organophosphorus pesticides and do not persist for as long. Figure 3.3 shows some of the basic structures for carbamate pesticides and Table 3.3 gives the common and chemical names for the compounds cited in the text.

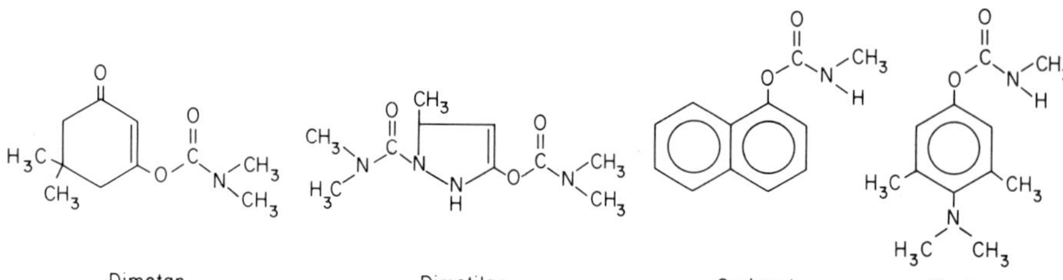

Fig. 3.3. Examples of dimethyl- and *N*-methyl-substituted carbamate insectides.

3.4.1 Thin layer chromatography

Finocchiaro and Benson[28] used Aluminum Oxide G and Silica Gel G-HR for separating some carbamate pesticides.

Adsorbent: Aluminum Oxide G

Solvent system: toluene–ethyl ether (3:1)

R_F: Hercules 9699 = 0.23; Hercules 8717 = 0.28; Baygon = 0.29; carbaryl = 0.33; MCA 600 = 0.34; UC 10854 = 0.43; RE 5305 = 0.44; Hooker 1422 = 0.50; dimetilan = 0.13; Isolan = 0.20; Pyrolan = 0.44; Azak = 0.52; 1-naphthyl *N,N*-dimethylcarbamate = 0.54.

The residues were rendered visible when a methanolic potassium hydroxide solution was sprayed onto the plate followed by *p*-nitrobenzenediazonium fluoborate solution which couples with the phenol formed by hydrolysis of the residue. Compounds without available coupling sites were detected by spraying with a solution of vanillin in sulfuric acid and then heating the plate.

Adsorbent: Silica Gel G-HR

Solvent system: toluene–ethyl ether (3:1)

R_F: Matacil = 0.15; Zectran = 0.34.

These compounds were made visible by spraying with a diphenylpicryl-hydrazyl solution and then heating the plate.

Abbott *et al.*[29] used thin layer chromatography to identify some carbamates in soil and water, with silica gel layers as the adsorbent.

Solvent system 1: hexane–acetone (9:1)

R_F: Propham = 0.60; Chlorpropham = 0.60; Barban = 0.40; carbaryl = 0.20; Arprocarb = 0.25.

Solvent system 2: hexane–acetone (7:3)

R_F: Propham = 0.77; Chlorpropham = 0.77; Barban = 0.65; carbaryl = 0.59; Arprocarb = 0.55.

Solvent system 3: chloroform

R_F: Propham = 0.72; Chlorpropham = 0.78; Barban = 0.80; carbaryl = 0.47; Arprocarb = 0.42.

Solvent system 4: methylene chloride

R_F: Propham = 0.86; Chlorpropham = 0.89; Barban = 0.86; carbaryl = 0.61; Arprocarb = 0.51.

TABLE 3.3
Common and chemical names of carbamate pesticides

Common name	Chemical name
Aldicarb	2-Methyl-2-(methylthio)propionaldehyde-*O*-(methylcarbamoyl)oxime
Aminocarb	4-(Diethylamino)-*m*-tolyl methylcarbamate
Arprocarb	See propoxur
Azak (terbutol)	2,6-Di-*tert*-butyl-*p*-tolyl methylcarbamate
Banol	See carbanolate
Baygon	See propoxur
Bendiocarb	2,3-(Isopropylidenedioxy)phenylmethylcarbamate
Benomyl	Methyl 1-(butylcarbamoyl)-2-benzimidazolecarbamate
Bufencarb	3-(1-Methylbutyl)phenyl methylcarbamate + 3-(1-ethylpropyl)phenyl carbamate (3:1)
Bux (metalkamate)	Mixed isomers of 3-*sec*-amylphenylmethylcarbamate
Carbanolate	6-Chloro-3,4-xylylmethylcarbamate
Carbaryl	1-Naphthyl methylcarbamate
Carbofuran	2,3-Dihydro-2,2-dimethyl-7-benzofuranylmethylcarbamate
Carbyne (barban)	4-Chlorobut-2-ynyl-3-chlorophenylcarbamate
Chloropropham	See CICP (organochlorine)
CPMC (Hopcide)	2-Chlorophenyl-*N*-methylcarbamate
Cycloate	*S*-Ethyl-*N*-ethylthiocyclohexanecarbamate
Dichlormate	3,4-Dichlorobenzyl methylcarbamate
Dimetan	5,5-Dimethyl-3-oxo-1-cyclohexen-1-yl dimethylcarbamate
Dimetilan	1-(Dimethylcarbamoyl)-5-methyl-3-pyrazolyl dimethylcarbamate
Dioxacarb	*o*-1,3-Dioxalan-2-ylphenyl methylcarbamate
Hercules 8717	3-Propargyloxyphenyl-*N*-methylcarbamate
Hercules 9699	2-Propargyloxyphenyl-*N*-methylcarbamate
Isolan	Dimethyl-5-(1-isopropyl-3-methylpyrazolyl)carbamate
Landrin	Mixture of 3,4,5-trimethylphenyl and 2,3,5-trimethylphenyl methylcarbamate
MCA-600	Benzo[*b*]thien-4-yl methylcarbamate
Matacil	See aminocarb
Meobal	3,4-Dimethylphenyl-*N*-methylcarbamate
Mesurol	See methiocarb
Methiocarb	4-(Methylthio)-3,5-xylyl methylcarbamate
Methomyl (lannate)	*S*-Methyl-*N*-(methylcarbamoyl oxy)thioacetimidate
Mexacarbate	4-(Dimethylamino)-3,5-xylyl methylcarbamate
Mobam	4-Benzothienyl-*N*-methylcarbamate
MTMC (Tsumacide)	*m*-Tolyl-*N*-methylcarbamate
Oxamyl	Methyl-*N′*,*N′*-dimethyl-*N*-[(methylcarbamoyl)oxy]-1-thiooxaimidate
Promecarb	*m*-Cym-5-yl methylcarbamate
Propham	See IPC (organochlorine)
Propoxur	*o*-Isopropoxyphenyl methylcarbamate
Pyrolan	Dimethyl-5-(3-methyl-1-phenylpyrazolyl)carbamate
Saturn (benthiocarb)	*S*-(4-Chlorobenzyl)-*N*,*N*-diethylthiolcarbamate
Solan (pentanochlor)	*N*-(3-Chloro-4-methylphenyl)-2-methylvaleramide
Stam (propanil)	3′,4′-Dichloropropionanilide
Swep	Methyl-*N*-(3,4-dichlorophenyl)carbamate
Temik	See aldicarb
UC 10854	*m*-Isopropylphenyl-N-methylcarbamate
Zectran (mexacarbate)	4-(Dimethylamino)-3,5-xylyl methylcarbamate

The compounds were detected by spraying with a 0.5% ethanolic solution of *p-N*-dimethyl benzaldehyde, then overspraying with a 2N aqueous solution of sulfuric acid and heating to 150 °C.

Mendoza and Shields[30] separated some carbamates on aluminum oxide.

Adsorbent: Aluminum Oxide DS-5 (Camag)

System A: pentane–benzene–acetone (7:1:2)

R_F: aldicarb = 0.60; Banol = 0.73; Baygon = 0.63; carbaryl = 0.57; carbofuran = 0.57; Matacil = 0.57; Mesurol = 0.63; Mobam = 0.52; Orthro 5353 = 0.73; Tanid = 0.03; Zectran = 0.73.

System B: hexane–acetone (8:2)

R_F: aldicarb = 0.20; Banol = 0.42; Baygon = 0.20; carbaryl = 0.20; carbofuran = 0.20; Matacil = 0.22; Mesurol = 0.27; Mobam = 0.16; Ortho 5353 = 0.30; Zectran = 0.30.

The migrated spots were made visible by spraying the plate with a serum[4] or a liver homogenate solution buffered at about pH 8 [Section 3.3.2].[4] The esterases in the serum or the liver homogenate were inhibited by carbamates. After a 30 min incubation period the plate is over-sprayed with one of the following substrates; indoxyl acetate, bromo-indoxyl acetate, naphthyl acetate or other similar esters. The indoxyl esters are hydrolyzed by the esterases and oxidize in air to form the blue dye, indigo. Where the enzymes are inactivated by carbamates, white spots are seen on a blue background. Other aromatic esters can couple with a diazonium salt to produce a dark background and reveal the carbamates as white or lighter colored spots against a darker background.

Getz[31] used precoated silica gel plates for resolving some carbamate pesticides.

Adsorbent: Merck precoated silica gel 20 × 20 cm (1968 vintage)

Solvent system: hexane–ethyl acetate–chloroform (70:20:5)

R_F: Dimetilan = 0.13; Temik = 0.33; Matacil = 0.48; carpofuran = 0.49; Baygon = 0.60; Mesurol = 0.65; Landin = 0.68; Zectran = 0.71; Methomyl = 0.73; Bux = 0.73; Banol = 0.73.

The carbamates were detected by dipping the plate into a 0.04% methanolic solution of 1,2,3-indantrione, air-drying it, and then heating at 110 °C for 10–15 min, or by dipping the plates into a 2% methanolic solution of sodium or potassium hydroxide, heating at 100 °C for 10–15 min and then dipping into a hexane solution of 0.1% 2,6-dichloroquinoneimine chloride. The aromatic carbamates became visible as blue spots and the sulfur-containing alkyl carbamates were seen as yellowish-brown spots.

The unusual hydrogen bonding qualities of polyamide were exploited by Nagasawa *et al.*[32] who used thin layers of it to resolve carbamate pesticides.

Adsorbent: Polyamide B-10 containing 10% of calcium sulfate

Solvent system I: cyclohexane–acetone (8:10)

R_F: carbaryl = 0.51; CPMC = 0.54; Baygon = 0.60; EMPC = 0.60; APC = 0.78; Meobal = 0.61; Cosban = 0.67; Osbac = 0.74;

MTMC = 0.63; Carbyne = 0.67; Swep = 0.60; methyl *N*-phenylcarbamate = 0.61; methyl *N*-α-naphthylcarbamate = 0.58; methyl *N*-cyclohexylcarbamate = 0.91; Saturn = 0.95; Carbac = 0.93; Solan = 0.72; Stam = 0.50.

Solvent system 2: methanol–water (5:5)

R_F: carbaryl = 0.42; CPMC = 0.62; Baygon = 0.73; EMPC = 0.52; APC = 0.32; Meobal = 0.60; Cosban = 0.58; Osbac = 0.58; MTMC = 0.67; Carbyne = 0.22; Swep = 0.20; methyl *N*-phenylcarbamate = 0.56; methyl *N*-α-naphthylcarbamate = 0.39; methyl *N*-cyclohexylcarbamate = 0.81; Saturn = 0.34; Carbac = 0.19; Solan = 0.21; Stam = 0.23.

Solvent system 3: acetone–water (4:6)

R_F: carbaryl = 0.57; CPMC = 0.67; Baygon = 0.74; EMPC = 0.59; APC = 0.40; Meobal = 0.65; Cosban = 0.65; Osbac = 0.59; MTMC = 0.70; Carbyne = 0.35; Swep = 0.34; methyl *N*-phenylcarbamate = 0.64; methyl *N*-α-naphthylcarbamate = 0.53; methyl *N*-cyclohexylcarbamate = 0.70; Saturn = 0.36; Carbac = 0.27; Solan = 0.32; Stam = 0.36.

Solvent system 4: *N*,*N*-dimethylformamide–water (4:6)

R_F: carbaryl = 0.62; CPMC = 0.72; Baygon = 0.80; EMPC = 0.61; APC = 0.35; Meobal = 0.71; Cosban = 0.70; Osbac = 0.60; MTMC = 0.74; Carbyne = 0.36; Swep = 0.30; methyl *N*-phenylcarbamate = 0.69; methyl *N*-α-naphthylcarbamate = 0.74; methyl *N*-cyclohexylcarbamate = 0.74; Saturn = 0.39; Carbac = 0.28; Solan = 0.30; Stam = 0.33.

The compounds were detected as dark spots on a light background when illuminated with black light. The plates were sprayed with a 1N ethanolic potassium hydroxide solution and then with a 10% ethanolic solution of ethylene glycol saturated with *p*-nitrobenezenediazonium fluoborate. *N*-Methyl carbamates with an open *ortho* or *para* position will react.

Treating the plate with bromine vapor and fluorescein will also detect the carbamates but involves a long wait for all the excess bromine to clear in order to reveal the carbamate.

The plates were sprayed with a solution of Rhodamine B (0.2 g dissolved in a mixture of 50 ml 0.02N silver nitrate and 50 ml of 1N hydrochloric acid), then dried and irradiated with UV radiation for 30 min.

3.5 ARSENICALS

Arsenic is still in use and thin layer chromatography has been used successfully for detecting some of its salts in soil. Von Endt *et al.*[33] used silica gel and cellulose for the separations.

Adsorbent: MN Cellulose 300G

Solvent: methanol–butanol–water (3:1:1)

R_F: monosodium methanearsonic acid = 0.60; sodium arsenite = 0.80; sodium arsenate = 0.55.

Adsorbent: Silica Gel G

Solvent: methanol–ammonium hydroxide–10% trichloracetic acid–water (50:15:5:30)

R_F: monosodium methanearsonic acid = 0.73; sodium arsenite = 0.57; sodium arsenate = 0.27–0.60.

Adsorbent: Silica Gel H

Solvent: as above

R_F: monosodium methanearsonic acid = 0.67; sodium arsenite = 0.45; sodium arsenate = 0.27–0.60.

3.6 HERBICIDES AND PLANT GROWTH REGULATORS

These chemicals can be used either to control weeds and to control or spread desirable plants. They include compounds with a range of chemical structures, as illustrated in Fig. 3.4. The chemical names of some common herbicides are

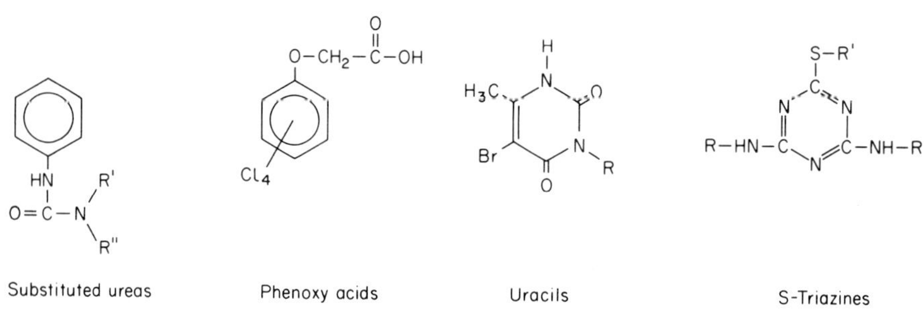

Fig. 3.4. Structural formulae showing the wide range of molecular structures that exhibit herbicidal activity.

listed in Table 3.4. The phenoxy herbicides are widely used for controlling weeds in crop stands, grazing land, lawns and for killing unwanted brush and trees. They are highly selective because they kill or control most broad-leaved (dicotyledonous) plants but do not affect the growth of grasses or grains (monocotyledonous plants). They are effective at low concentrations which minimizes any effects on man, domestic animals or game.

Symmetrical triazines control a wide range of unwanted vegetation. Their selectivity depends upon the substituent groups attached to the triazine nucleus.

Other herbicides contain carbamate, substituted urea, nitroaniline and chloroalkyl acid nucleii.

TABLE 3.4
Common and chemical names of herbicides

Common name	Chemical name
Ametryn	2-(Ethylamino)-4-(isopropylamino)-6-(methylthio)-*s*-triazine
Amitrol	3-Amino-*s*-triazole
Anilazine	2,4-Dichloro-6-(*o*-chloroanilino)-*s*-triazine
Atraton	2-(Ethylamino)-4-(isopropylamino)-6-methoxy-*s*-triazine
Atrazine	2-Chloro-4-(ethylamino)-6-(isopropylamino)-*s*-triazine
Benzthiazuron	1-(2-Benzothiazolyl)-3-methylurea
Bromacil	5-Bromo-3-*sec*-butyl-6-methyluracil
Buturon	3-(*p*-Chlorophenyl)-1-methyl-1-(1-methyl-2-propynyl)urea
Chlorazine	2-Chloro-4,6-bis(diethylamino)-*s*-triazine
Chlorbromuron	3-(4-Bromo-3-chlorophenyl)-1-methoxy-1-methylurea
Chloroxuron	3-[*p*-(*p*-Chlorophenoxy)phenyl]-1,1-dimethylurea
3-CP	2-(3-Chlorophenoxy)propionic acid
3-CPA	2-(3-Chlorophenoxy)propionamide
4-CPA	(4-Chlorophenoxy)acetic acid
Cycluron	3-Cyclooctyl-1,1-dimethylurea
Dalapon	2,2-Dichloropropionic acid
2,4-D	(2,4-Dichlorophenoxy)acetic acid
2,4-DB	4-(2,4-Dichlorophenoxy)butyric acid
Desmetryn	2-(Isopropylamino)-4-(methylamino)-6-(methylthio)-*s*-triazine
Dichlorprop	2-(2,4-Dichlorophenoxy)propionic acid
Dipropetryn	2-(Ethylthio)-4,6-bis(isopropylamino)-*s*-triazine
Diuron	3-(3,4-Dichlorophenyl)-1,1-dimethylurea
DNAP (dinosam)	2-(1-Methylbutyl)-4,6-dinitrophenol
DNBP (dinoseb)	2-*sec*-Butyl-4,6-dinitrophenol
DNOC	Dinitrocresol
DNOCHP (dinex)	2,4-Dinitro-6-cyclohexyl phenol
Fenac	(2,3,6-Trichlorophenyl)acetic acid
Fenuron	1,1-Dimethyl-3-phenylurea
Fluometuron	1,1-Dimethyl-3-(3-trifluoro methylphenyl) urea
Ipazine	2-Chloro-4-diethylamino-6-isopropylamino-*s*-triazine
Isocil	5-Bromo-3-isopropyl-6-methyluracil
Lenacil	3-Cyclohexyl-5,6-trimethyleneuracil
Linuron	3-(3,4-Dichlorophenyl)-1-methoxy-1-methylurea
MCPA	[(4-Chloro-*o*-tolyl)oxy]acetic acid
MCPB	4-[(4-Chloro-*o*-tolyl)oxy]butyric acid
Metobromuron	3-(*p*-Bromophenyl)-1-methoxy-1-methylurea
Methabenzthiazuron	1-(2-Benzothiazolyl)-1,3-dimethylurea
Methoprotryne	2-Isopropylamino-4-(3-methoxypropylamino)-6-methylthio-*s*-triazine
Monolinuron	3-(*p*-Chlorophenyl)-1-methoxy-1-methylurea
Monuron	3-(*p*-Chlorophenyl)-1,1-dimethylurea
Neburon	1-*N*-Butyl-3-(3,4-dichloro-phenyl)-1-methylurea
Norazine	2-Chloro-4-isopropylamino-6-methylamino-*s*-triazine
Norea	3-Hexahydro-4,7-methanoindan-5-yl)-1,1-dimethylurea
PCP	Pentachlorophenol
Prometon	2-Methoxy-4,6-bis(isopropylamino)-*s*-triazine
Prometryne	2-Methylthio-4,6-bis(isopropylamino)-*s*-triazine
Propazine	2-Chloro-4,6-bis(isopropylamino)-*s*-triazine
Siduron	1-(2-Methylcyclohexyl)-3-phenylurea

TABLE 3.4—*cont.*

Common name	Chemical name
Silvex	2-(2,4,5-Trichlorophenoxy)propionic acid
Simazine	2-Chloro-4,6-bis(ethylamino)-*s*-triazine
Simeton	2,4-bis(Ethylamino)-6-methoxy-*s*-triazine
Simetryn	2,4-bis(Ethylamino)-6-(methylthio)-*s*-triazine
2,4,5-T	(2,4,5-Trichlorophenoxy)acetic acid
2,3,6-TBA	2,3,6-Trichlorobenzoic acid
Tebuthiuron	*N*-[5-(1,1-Dimethylethyl)-1,3,4-thidiazol-2-yl]-*N*,*N'*-dimethylurea
Terbacil	3-*tert*-Butyl-5-chloro-6-methyluracil
Terbutryn	2-(*tert*-Butylamino)-4-(ethylamino)-6-(methylthio)-*s*-triazine
Trietazine	2-Chloro-4-diethylamino-6-ethylamino-*s*-triazine
Trimeturon	1-(*p*-Chlorophenyl)-2,3,3-trimethyl*pseudo*urea

3.6.1 Extraction and clean-up

The highly polar nature of the phenoxy herbicides necessitates the use of a different type of extraction. Abbott *et al.*[34] acidified a 100 ml aqueous sample of herbicide with 5 ml of 6N sulfuric acid and then extracted with two 100 ml portions of ethyl ether. The ethyl ether extracts were combined and then partitioned twice with 150 ml of 8% sodium sulfate solution containing 2.5N sodium hydroxide. The new aqueous extracts were acidified with 10 ml of 6N sulfuric acid and then extracted with 150 ml and 100 ml aliquots of ethyl ether respectively. The ether extracts were dried by filtering through sodium sulfate, and then evaporated and the residue taken up with 40 μl of ethyl acetate for spotting on the chromatogram.

A dry soil sample was extracted by grinding with a small volume of water in a mortar. A 100 g sample of this mixture was diluted with a further 100 ml of water and the slurry transferred to a flask. Twenty ml of 6N sulfuric acid and 200 ml of ethyl ether were added to the slurry and any herbicide residue present was extracted by tumbling the mixture for 1 h. The contents of the flask were centrifuged, and the solvent layer was decanted and treated in the same way as the water extract.

Whatman No. 3 paper was used in combination with a mobile phase of paraffin, benzene, glacial acetic acid and cyclohexane (10:30:20:200). The following R_F values were obtained: MCPB = 0.81; 2,4-DB = 0.79; MCPA = 0.68; 2,4,5-T = 0.66; 2,4-D = 0.58.

The herbicides were detected by spraying the chromatogram with 0.5% silver nitrate in 95% ethanol, exposing it to high intensity UV radiation, and then treating it with 0.5% sodium thiosulfate. The excess thiosulphate was washed out by soaking in water. The chromatogram was then dried in an oven at 80 °C.

Yip[35] extracted herbicide residue from green vegetables by acidifying a 100 g fresh sample with 10 ml of 10% sulfuric acid and then homogenizing in 25 ml of ethanol, 50 ml of petroleum ether and 150 ml of ethyl ether for 3 min. The mixture was centrifuged and the supernatant solvent decanted through sodium sulfate into a centrifuge bottle. One hundred ml of 3% sodium bicarbonate were added to the decanted solvent, and the container was shaken and then centrifuged. The top layer was removed and a further 25 ml of ethyl ether added and the centrifuging repeated.

The aqueous solution contained the free acids and the ether layer contained the esters. The free acids were extracted into chloroform and then methylated with diazomethane. Both the aqueous and ether extracts were then ready for chromatography.

Wheat was extracted with 200 ml of 95% ethanol for each 100 g of sample; the ethanol was evaporated and the residue taken up with chloroform and converted to the methyl ester with diazomethane. The herbicides were then identified using paper chromatography.

3.6.2 Paper chromatography

Adsorbent: Whatman No. 1 paper
System A: Immobile phase: 10% mineral oil in ethyl ether
Mobile phase: 50% acetonitrile in water

R_F: Silvex methyl ester = 0.31; 2,4-DB methyl ester = 0.34; 2,4,5-T isopropyl ester = 0.48; MCP isopropyl ester = 0.50; 2,4-D-isopropyl ester = 0.64.

System B: Immobile phase: 35% *N,N*-dimethylformamide in ethyl ether
Mobile phase: 2,2,4-trimethylpentane

R_F: Silvex methyl ester = 0.82; 2,4-DB methyl ester = 0.73; 2,4,5-T isopropyl ester = 0.50; MCP isopropyl ester = 0.58; 2,4-D isopropyl ester = 0.39.

Mitchell[36] separated some urea herbicides by paper chromatography.

Adsorbent: Whatman No. 1 paper
Immobile phase: 10% heavy mineral oil in ethyl ether
Mobile phase: tetrahydrofuran–water (25:75)

R_F: fenuron = 0.93; monuron = 0.78; diuron = 0.50; 3-(2-chlorophenyl)-1-methylurea = 0.81; linuron = 0.18; neburon = 0.23.

Mobile phase: 2,2,4-trimethylpentane–acetic acid (75:25)

R_F: fenuron = 0.21; monuron = 0.27; diuron = 0.33; 3-(2-chlorophenyl)-l-methylurea = 0.30; linuron (streaks); neburon = 0.79.

The compounds were detected with silver nitrate–phenoxyethanol reagent.

3.6.3 Thin layer chromatography

Aluminum oxide has been used for resolving methyl esters of the phenoxy herbicides.[37] The developing solvent was hexane saturated with acetonitrile.

The following R_F values were found: 2,4-D = 0.20; 2,4,5-T = 0.23; 2,4-DB = 0.30; MCPA = 0.30; 2,4,5-TP = 0.41; 2,3,6-TBA = 0.51; PCP = 0.73.

The free acids were separated on Eastman silica sheets.[37] The sheets were developed first in cyclohexane and acetic acid (10:1) for a distance of 3.5 inches then removed, dried and then developed again in a new tank containing benzene and petroleum ether (3:1) as the developing solvents. R_F values obtained were: 2,4-D = 0.28; 2,4,5-T = 0.33; MCPA = 0.44; 2,4,5-TP = 0.59; 2,3,6-TBA = 0.64, 0.42, 0.20; 2,4-DB = 0.68; PCP = 0.87. The compounds were detected with silver nitrate–phenoxyethanol reagent.

Yip and Howard[38] coated plates with cellulose powder (MN 300HR) and applied immobile–mobile phases for the separation of some dinitrophenol compounds.

Immobile phase: mineral oil–acetic acid–ethyl ether (5:2:93)

Mobile phase: methanol–acetic acid–water (73:2:25)

R_F: DNOC = 0.83; DNBP = 0.65; DNAP = 0.52; DNOCHP = 0.44; 2,4-dinitrophenol = 0.85; *p*-nitrophenol = 0.90.

The compounds were detected by reducing the nitro groups with a hydrochloric acid solution of stannous chloride (3 ml of 15% stannous chloride + 180 ml of water + 15 ml of concentrated hydrochloric acid). The plate was sprayed with the solution, allowed to dry and then over-sprayed with Ehrlich's reagent (1 g of *p*-dimethylaminobenzaldehyde + 30 ml of ethanol + 3 ml of concentrated hydrochloric acid + 180 ml of *n*-butanol). The dinitro compounds appeared as yellow-orange spots which fluoresced under UV radiation.

Onley and Yip[39] used several solvent systems with a layer of Aluminum Oxide GF.

System A: benzene–butyl ether–dioxane (40:20:1)

R_F: 3,4-dichloroaniline = 0.81; *p*-chloroaniline = 0.77; linuron = 0.73; metabromuron 0.71; neburon = 0.62; fluometuron = 0.40; diuron = 0.38; chloroxuron = 0.32; monuron = 0.31; 3,3,4-dichlorophenyl-l-methoxyurea = 0.27; fenuron = 0.22; 3,3,4-dichlorophenyl-l-methylurea = 0.11; 3-*p*-chlorophenyl-l-methylurea = 0.09; 3,3,4-dichlorophenylurea = 0.02; 3-*p*-chlorophenylurea = 0.01.

System B: benzene–chloroform–methanol (25:9:1)

R_F: 3,4-dichloroaniline = 0.71; *p*-chloroaniline = 0.73; linuron = 0.81; metabromuron = 0.83; neburon = 0.72; fluometuron = 0.55; diuron = 0.54; chloroxuron = 0.64; monuron = 0.54; 3,3,4-dichlorophenyl-l-methoxyurea = 0.39; fenuron = 0.52; 3,3,4-dichlorophenylurea = 0.02; 3-*p*-chlorophenylurea = 0.06.

System C: benzene–butyl ether–methylene chloride–dioxane (26:10:10:1)

R_F: 3,4-dichloroaniline = 0.67; *p*-chloroaniline = 0.60; linuron = 0.60; metabromuron = 0.65; neburon = 0.50; fluometuron = 0.38; diuron = 0.28; chloroxuron = 0.31; monuron = 0.24; 3,3,4-dichlorophenyl-l-methoxyurea = 0.18; fenuron = 0.17; 3,3,4-

dichlorophenyl-l-methylurea = 0.08; 3-*p*-chlorophenyl-l-methylurea = 0.08; 3,3,4-dichlorophenylurea = 0.01; 3-*p*-chlorophenylurea = 0.01.

The migrated compounds were detected by spraying the plate with a 5% potassium hydroxide solution, and then heating at 190 °C for 20 min. After cooling, the plate was sprayed with 20 ml of concentrated hydrochloric acid dissolved in 100 ml of methanol. The acid was allowed to react for 5 min and the plate then sprayed with a 1% sodium nitrite solution and with 1 g of *N*-(l-naphthyl)ethylenediamine dissolved in water and diluted to 100 ml in methanol. Reddish-purple spots appeared.

Abbott *et al.*[34] resolved some herbicides by using the same developing solvent but mixing different ratios of silica gel and kieselguhr.

Developing solvent: cyclohexane–benzene–acetic acid–liquid paraffin (20:3:2:1)

Adsorbent: Silica Gel G

R_F: MCPB = 0.18; 2,4-DB = 0.17; MCPA = 0.14; 2,4,5-T = 0.13; 2,4-D = 0.12; dalapon = 0.10.

Adsorbent: 30% Kieselguhr G + 70% Silica Gel G

R_F: MCPB = 0.52; 2,4-DB = 0.44; MCPA = 0.26; 2,4,5-T = 0.20; 2,4,D = 0.18; dalapon = 0.17.

Adsorbent: 40% Kieselguhr G + 60% Silica Gel G

R_F: MCPB = 0.61; 2,4-DB = 0.52; MCPA = 0.32; 2,4,5-TP = 0.25; 2,4-D = 0.21; dalapon = 0.20.

The spots were rendered visible with the silver nitrate–phenoxyethanol reagent.

Henkel and Ebing[40] used Silica Gel G to separate some of the triazines.

Adsorbent: Silica Gel G

Solvent 1: chloroform–nitromethane (1:1)

R_F: simeton = 0.26; atraton = 0.34; prometon = 0.42; simetryne = 0.50; ametryne = 0.59; prometryne = 0.68.

Solvent 2: chloroform + nitromethane (5:1)

R_F: simazine = 0.28; atrazine = 0.37; propazine = 0.48; trietazine = 0.60; ipazine = 0.66; chlorazine = 0.80.

The compounds were detected by spraying with 0.1N aqueous silver nitrate solution and subsequent UV irradiation; brownish spots were formed where triazines were present.

Ebing[41] developed an elaborate system for identifying triazines, substituted ureas and uracil compounds.

Adsorbent: precoated silica gel with aluminum backing. Before chromatography, the humidity of the adsorbent was equilibrated with that of a solution of sulfuric acid with a density of 1.6

Solvent: isopropyl ether

R_F: norazin = 0.15; simazine = 0.17; propazine = 0.31; trietazine = 0.54; atrazine = 0.25; ipazine = 0.62; chlorazine = 0.85; anilazine =

0.71; simeton = 0.03; atraton = 0.04; prometon = 0.06; *nor*-ametryne = 0.14; simetryne = 0.15; ametryne = 0.22; GS 14260 = 0.29; prometryne = 0.30; methoprotryne = 0.06; amitrol = 0.06.

Alternatively the same adsorbent was equilibrated with an aqueous sulfuric acid solution of density 1.35.

Solvent: nitromethane

R_F: norazin = 0.58; simazine = 0.52; propazine = 0.73; trietazine = 0.85; atrazine = 0.65; ipazine = 0.92; chlorazine = 1; anilazine = 1; simeton = 0.21; atraton = 0.27; prometon = 0.34; *nor*ametryne = 0.53; simetryne = 0.53; ametryne = 0.66; GS 14260 = 0.75; prometryne = 0.76; methoprotryne = 0.34; amitrol = 0.34.

The compounds were rendered visible with silver nitrate solution and subsequent UV irradiation.

Adsorbent: precoated Silica Gel DC with aluminum support, the adsorbent was equilibrated with a sulfuric acid solution of density 1.35, and the plates developed in a sandwich chamber

Solvent: cyclohexane–ethyl acetate (1:1)

R_F: norea = 0.30; benzthiazuron = 0.31; methabenzthiazuron = 0.54; fenuron = 0.34; monuron = 0.34; monolinuron = 0.84; diuron = 0.43; linuron = 0.90; neburon = 0.95; metabromuron = 0.88; fluometuron = 0.47; chlorbromuron = 0.88; cycluron = 0.27; chloroxuron = 0.36; buturon = 0.95; siduron = 0.96; Benzomarc = 0.77; Metoxymarc – 0.58.

The compounds were detected by spraying with a 1% solution of *p*-(*N*,*N*-dimethylamino)benzaldehyde in ethanol with 5 ml of concentrated hydrochloric acid. The plate required heating at 200 °C for 30 min before spots became visible.

Compounds which would not react with this reagent were detected by spraying the plate with a 0.005% aqueous solution of Rhodamin B and then viewing under filtered UV (black light).

Adsorbent: precoated Polyamide DC on aluminum plates, humidity regulated by equilibrating in aqueous sulfuric acid with a density of 1.35

Solvent: cyclohexane–dichloromethane (3:1)

R_F: methabenzthiazuron = 0.49; fenuron = 0.19; monuron = 0.10; monolinuron = 0.56; diuron = 0.06; linuron = 0.46; neburon = 0.15; metabromuron = 0.56; fluometuron = 0.13; chlorbromuron = 0.44; chloroxuron = 0.21; buturon = 0.17; siduron = 0.02; Benzomarc = 0.82; Metoxymarc = 0.78.

Detection was carried out by viewing the chromatogram under filtered UV radiation at 254 nm, or by spraying with Rhodamin B solution.

Adsorbent: Silica Gel G (2.8 g) + Silica Gel H (particle size less than 20 μm) 0.7 g suspended in 14 ml for coating one 20 × 20 cm plate. The plates were equilibrated with a sulfuric acid solution of density 1.6. Two analytical plates were placed together for sandwich development

Solvent: nitromethane

R_F: norea = 0.10; benzthiazuron = 0.24; methabenzthiazuron = 0.52; fenuron = 0.28; monuron = 0.37; monolinuron = 0.86; diuron = 0.57; linuron = 0.88; neburon = 0.85; metabromuron = 0.84; fluometuron = 0.58; chlorbromuron = 0.92; cycluron = 0.08; chloroxuron = 0.39; buturon = 0.92; siduron = 0.79; Benzomarc = 0.85; Metoxymarc = 0.82.

The compounds were rendered visible by spraying with a 3% ethanolic solution of ninhydrin containing 5 ml of acetic acid, and then heating at 180 °C for 30 min. Several spots became visible as a result, though reheating at 150 °C was required to detect other compounds.

Adsorbent: precoated Silica Gel-DC on aluminum backing, equilibrated with sulfuric acid of density 1.6

Solvent: toluene–ethyl acetate (3:1)

R_F: isocil = 0.21; bromacil = 0.25; terbacil = 0.30; lenacil = 0.15.

The compounds were detected by spraying with a 0.5% solution of Malachite Green in acetone and then treating with bromine vapor.

Adsorbent: 1 g Polyamide-DC with 0.02 g of fluorescent indicator suspended in 9.5 ml of absolute ethanol (sufficient for coating one plate), equilibrated with sulfuric acid of density 1.35

Solvent: 33% aqueous ethanol

R_F: isocil = 0.47; bromacil = 0.40; terbacil = 0.38; lenacil = 0.33.

Compounds were detected by their quenching effect under filtered UV illumination, or by spraying with a 0.01% aqueous ethanolic solution of Rhodamine B.

3.7 FUNGICIDES

Fungicides are important because they help to reduce the spoilage of food during storage or transportation over long distances. In addition to the problem of spoilage, there is growing concern about the formation of mycotoxins on food crops while they are being grown or stored. The fungicides which have so far been analyzed by paper and thin layer chromatography, are the mercury-containing compounds, the dithiocarbamates, and the organochlorine compounds.

Tatton and Wagstaffe[42] analyzed the mercury-containing compounds by converting them to the dithizonates and then chromatographing these on silica gel and alumina. The dithizonate compounds, with their red or yellow colors, are self-indicators.

Adsorbent: silica gel

Solvent: hexane–acetone (9:1)

R_F: methylmercury = 0.64; ethylmercury = 0.64; methoxyethylmercury = 0.32; ethoxyethylmercury = 0.44; phenylmercury = 0.48; tolylmercury = 0.52; mercury dithizonate = 0.19.

Solvent 2: hexane–acetone (93:7)

R_F: methylmercury = 0.57; ethylmercury = 0.62; methoxyethylmercury = 0.25; ethoxyethylmercury = 0.34; phenylmercury = 0.46; tolylmercury = 0.53; mercury dithizonate = 0.17.

Solvent 3: petroleum ether–acetone (9:1)

R_F: methylmercury = 0.77; ethylmercury = 0.78; methoxyethylmercury = 0.44; ethoxyethylmercury = 0.55; phenylmercury = 0.62; tolylmercury = 0.69; mercury dithizonate = 0.28.

Adsorbent: alumina

Solvent: hexane–acetone (19:1)

R_F: methylmercury = 0.89; ethylmercury = 0.91; methoxyethylmercury = 0.58; ethoxyethylmercury = 0.71; phenylmercury = 0.72; tolylmercury = 0.79; mercury dithizonate = 0.19.

Geike and Schupan[43] separated the organomercury fungicides on thin layers of silica gel and characterized them by their inhibitory effects on several enzyme systems.

Adsorbent: Silica Gel G or Silica Gel GF

Solvent A: benzene–acetone (90:10)

R_F: methylmercury chloride = 0.79; methylmercury sulfate = 0.0; methoxyethylmercury chloride = 0.75, 0.50, 0.35. 0.18; phenylmercuryacetate = 0.97, 0.77; methylmercury mercaptobenzoate = 0.77; phenylmercury benzcatechin = 0.95, 0.78, 0.29; methylmercury hydroxyquinolate = 0.75, 0.58, 0.25; methylmercury cyanoguanidine = 0.74; methylmercurychlorodiacetamide = 0.85, 0.50.

These compounds were also detected by their inhibitory effect on enzymes. The plate was sprayed with a solution of urease in a pH 7 buffer. After incubating with the enzyme, the substrate containing 500 mg of urea and 12.5 ml of 0.01 % Bromthymol Blue in a pH 7 buffer was sprayed onto the plate. Where mercury compounds were present, there was no hydrolysis of the urea to ammonia so that the dye retained its yellow color at pH 7. Where mercury was present, ammonia was released, which raised the pH and changed the dye to its basic blue color in that area.

Hylin[44] separated the thiocarbamates on Silica Gel G using two different solvent systems.

Solvent 1: benzene

R_F: thiuram = 0.17; tetramethylthiuram monosulfide = 0.30; ziram = 0.68; zineb = 0.95; maneb = 0.98.

Solvent 2: benzene–methanol–acetic acid (48:8:4)

R_F: thiuram = 0.86; zineb = 0.88; ziram = 0.94; maneb = 0.98.

The compounds were made visible by treating the plate with a solution of 0.5 % cuprous chloride, 1.5 % ammonium chloride, 1.5 ml of ammonium hydroxide and 15 % hydroxylamine hydrochloride.

REFERENCES

1. R. Carson, *Silent Spring*, Houghton Miflin, Boston, Mass, 1962.
2. R. von Müller, G. Ernst and H. Schoch, *Mitt. Geb. Lebensmittelunters. Hyg.* **48**, 152 (1957).
3. L. R. Jones and J. A. Riddick, *Anal. Chem.* **24**, 569 (1952).
4. A. K. Klein, *J. Assoc. Off. Agric. Chem.* **42**, 539 (1959).
5. R. O. Mumma, W. B. Wheeler, D. E. H. Frear and R. H. Hamilton, *Science* **152**, 530 (1966).
6. M. E. Getz, *J. Assoc. Off. Agric. Chem.* **45**, 393 and 397 (1962).
7. P. A. Mills, *J. Assoc. Off. Agric. Chem.* **43**, 81 (1960).
8. M. J. de Faubert Maunder, H. Egan, E. W. Godly, E. W. Hammond, J. Roburn and J. Thompson, *Analyst* **89**, 168 (1964).
9. T. Olson, Chemagricultural Corporation Report No. 10 081, 1962.
10. M. Eidelman, *J. Assoc. Off. Agric Chem.* **45**, 672 (1962); **46**, 182 (1963).
11. L. Kahn and C. H. Wayman, *Anal. Chem.* **36**, 1340 (1964).
12. J. I. Teasley and W. S. Cox, *J. Am. Water Works Assoc.* **55**, 1093 (1963).
13. L. C. Mitchell, *J. Assoc. Off. Agric. Chem.* **41**, 781 (1958).
14. W. P. McKinley and J. H. Mahon, *J. Assoc. Off. Agric. Chem.* **42**, 725 (1958).
15. P. A. Mills, *J. Assoc. Off. Agric. Chem.* **44**, 171 (1961).
16. K. C. Walker and M. Beroza, *J. Assoc. Off. Agric. Chem.* **46**, 250 (1963).
17. M. K. Brandt, Analtech Inc., Project No. AS-60-11-F., February 7, 1967.
18. M. F. Kovacs, Jr, *J. Assoc. Off. Anal. Chem.* **49**, 635 (1966).
19. E. J. Thomas, J. A. Burje and J. H. Lawrence, *J. Chromatogr.* **35**, 119 (1968).
20. L. C. Mitchell, *J. Assoc. Off. Agric. Chem.* **43**, 810 (1960).
21. H. F. MacRae and W. P. McKinley, *J. Assoc. Off. Agric. Chem.* **44**, 207 (1961).
22. J. A. R. Bates, *Analyst* **90**, 453 (1965).
23. M. F. Kovacs, Jr, *J. Assoc. Off. Agric. Chem.* **47**, 1097 (1964).
24. M. E. Getz and H. G. Wheeler, *J. Assoc. Off. Anal. Chem.* **51**, 1101 (1968).
25. D. C. Abbott, A. S. Burridge, J. Thomson and K. S. Webb, *Analyst* **92**, 170 (1967).
26. M. E. Getz, *J. Assoc. Off. Agric. Chem.* **47**, 1103 (1964).
27. R. C. Blinn, *J. Assoc. Off. Agric. Chem.* **46**, 952 (1963).
28. J. M. Finocchiaro and W. R. Benson, *J. Assoc. Off. Anal. Chem.* **50**, 888 (1967).
29. D. C. Abbott, K. W. Blake, K. R. Tarrant and J. Thomson, *J. Chromatogr.* **30**, 136 (1964).
30. C. E. Mendoza and J. B. Shields, *J. Chromatogr.* **50**, 92 (1970).
31. M. E. Getz, *Methods in Residue Analysis*, Ed. by A. S. Tahori. Gordon and Breach, London, 1971, pp. 57–58.
32. K. Nagasawa, H. Yoshidome and F. Kamata, *J. Chromatogr.* **52**, 453 (1970).
33. D. W. von Endt, P. C. Kearney and D. D. Kaufman, *J. Agric. Food Chem.* **16**, 17 (1968).
34. D. C. Abbott, H. Egan, E. W. Hammond and J. Thomson, *Analyst* **89**, 480 (1964).
35. G. Yip, *J. Assoc. Off. Agric. Chem.* **45**, 367 (1962).
36. L. C. Mitchell, *J. Assoc. Off. Anal. Chem.* **49**, 1163 (1966).
37. *Pesticide Analytical Manual*, Sections 421.4 and 422.4, US Dept of HEW, Food and Drug Administration, 1969.
38. G. Yip and S. F. Howard, *J. Assoc. Off. Anal. Chem.* **49**, 1166 (1966).
39. J. H. Onley and G. Yip, *J. Assoc. Off. Anal. Chem.* **52**, 526 (1969).
40. H. G. Henkel and W. Ebing, *J. Chromatogr.* **14**, 283 (1964).
41. W. Ebing, *J. Chromatogr.* **65**, 533 (1972).
42. J. O'G. Tatton and P. J. Wagstaffe, *J. Chromatogr.* **44**, 284 (1969).
43. F. Geike and I. Schupan, *J. Chromatogr.* **72**, 153 (1972).
44. J. W. Hylin, *Bull. Environ. Contam. Toxicol.* **1**, 77 (1966).

CHAPTER 4

FOOD ADDITIVES

The commercial preparation of food products has become a thriving industry in the economically developed countries where most of the population can afford a good diet.

Whereas early man used processes such as drying, salting, smoking and fermentation to preserve food and thus extend his food supply, in the present day they are replaced or supplemented by canning, freezing, refrigeration and dehydration, in conjunction with the use of chemical preservatives. Some chemical additives may also, of course, be used to enhance the aesthetic qualities of food.

Food additives are defined by the National Academy of Sciences as those chemicals which may be incorporated into foodstuffs, directly or indirectly while they are being grown, stored or processed. Chemical additives are therefore important environmental constituents which are regularly ingested by the consumer. As a result government agencies have been set up to regulate and supervise the use of such compounds, both to ensure a fair deal for the consumer and to prevent the use of toxic materials in food processing.

In the United States the Food and Drug Administration is the agency designated by law to oversee the use of food additives. In simple terms, their task is to determine the maximum amount of additive allowable in food, and also to conduct a constant survey of relevant new laboratory data in order to detect any possible hazards to the consumer.

Paper and thin layer chromatography, in conjunction with suitable isolation procedures, are among the methods used by research workers for detecting many of these compounds.

4.1 ANTIOXIDANTS AND PRESERVATIVES

When food oxidizes its fat and carbohydrate components break down and in turn cause the deterioration of proteins. Some results of oxidation are the rancidification of fats, the growth of moulds on carbohydrates and the putrefaction of proteins.

All organic foods contain microorganisms which live within or on the surface of larger organisms. Most common among these are bacteria, moulds and yeasts, some of which are useful to man and others of which are harmful. Raw fruits and vegetables also contain enzymes which accelerate their ripening processes; these processes need to be inhibited in order to lengthen the storage life of such foods, and preservatives are generally used for this purpose.

Antioxidants, as indicated by their name, are used as inhibitors of the oxidation process in food. They may be incorporated in the food product, or the food container may be impregnated with them. Polyvalent phenols, especially those derived from gallic acid, are a particularly useful group of antioxidants for foods with an acid pH value. The most widely used of these is *nor*hydroguiaretic acid (NDGA) [4,4′-(2,3-dimethyl tetramethylene)dipyrocatechol]. Its preservative effect is high in pH range 5–7, but poor in alkaline conditions. Butylated hydroxytoluene (BHT) (2,6-di-*tert*-butyl-4-methylphenol) is a widely used antioxidant for animal and vegetable fats. Butylate hydroxyanisole (BHA) (2-*tert*-butyl-4-methoxyphenol) is also an important antioxidant because it is effective at high (alkaline) pH values; it is generally used as a mixture of the two isomers, 2-BHA and 3-BHA.

Other antioxidants include ethoxyquin, esters of gallic acid, ascorbic acid and tocopherol. Figures 4.1 and 4.2 show the structures of some antioxidants and preservatives, respectively.

4.1.1 Isolation procedures

Ter Heide[1] dissolved a sample of fat or oil in a hydrocarbon such as hexane or petroleum ether. The solution of hydrocarbon and fat was then shaken in a

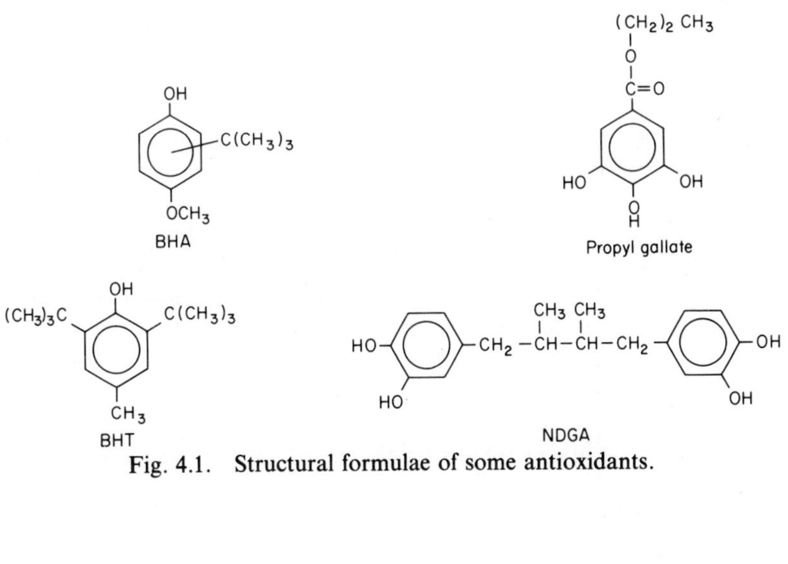

Fig. 4.1. Structural formulae of some antioxidants.

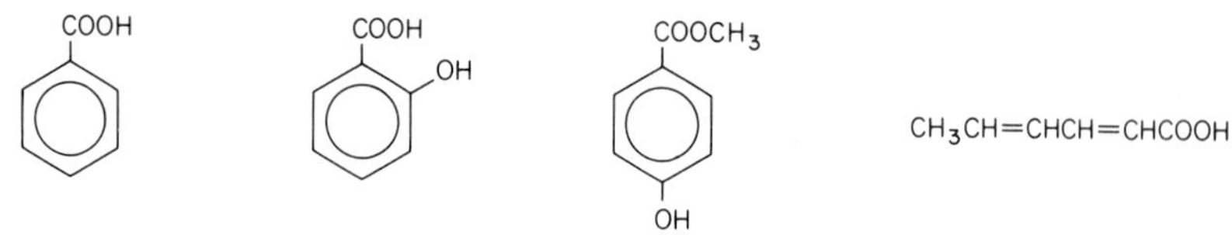

Benzoic acid Salicylic acid Methyl *p*-hydroxy benzoate Sorbic acid

Fig. 4.2. Structural formulae of some preservatives.

separating funnel with an equal volume of methanol, and enough water added to it to constitute 20% of the total volume. The separating funnel was shaken to form two separate layers, one of aqueous methanol and one of the hydrocarbon. The methanol layer was drawn off and concentrated by evaporation on a steam bath (or in a rotary evaporator) at 40 °C, and the resulting concentrate cooled to 0 °C in order to precipitate some of the fat still soluble in the aqueous methanol. The cooled solution was filtered and the filtrate extracted with ethyl acetate. The ethyl acetate extract was dried over sodium sulfate, after which it was ready for spotting on a paper or thin layer chromatogram.

However, BHT could not be recovered by this procedure, and a modification of Seher's method[2] was necessary in order to extract it. The fat remaining in the hydrocarbon layer after partitioning with aqueous methanol was saponified with alcoholic potassium hydroxide and the BHT, which was now in the alcoholic potassium hydroxide layer, could be isolated by drawing off this layer, acidifying it, and partitioning back into the hydrocarbon again.

Philips and Hinkel[3] dissolved the fat or oil in cyclohexane and then chromatographed the solution on a column containing 30 g of silicic acid. The cyclohexane eluent was collected after each 50 ml aliquot had flowed through the column. Other silica gels could be used, but in such cases it would have to be determined experimentally whether the eluate retained the antioxidant or not.

Scheidt and Conroy[4] dissolved the fat or oil in a hydrocarbon such as hexane, placed the solution in a separating funnel and then partitioned with four 25 ml aliquots of acetonitrile saturated with the hydrocarbon.

The partitioning procedure was further developed by Schwien *et al.*,[5] who evaporated the acetonitrile and took up the oily residue with 20 g of acid alumina for every 15 g of fat or oil. The adsorbent and oil were transferred to a column containing 5 g of the acid alumina and then eluted with 25% water in methanol. An eluate of 150 ml was collected, transferred to a separating funnel and extracted with ethyl ether and petroleum ether (1:1) and 10–15 ml of saturated sodium chloride solution. The organic layer was drawn off and concentrated for spotting on a chromatogram.

McCauley *et al.*[6] were able to isolate antioxidants from lard by using short-path vacuum distillation adapted from a sublimation apparatus. Four grams of sample were placed in a 60 ml flat-bottomed distillation flask and connected to the sublimation apparatus. The vacuum applied lowered the pressure to 0.5 microns. The distillation flask was kept in an insulated bath which had a temperature of 188 °C, and the cold finger used for collection was cooled with liquid nitrogen. The distillation was run for 2 h. The collected antioxidants in the cold finger were then washed out with ethyl ether into a clean container, after which the sample was ready for analysis.

Liquids or slurries of solids were steam distilled by Pinella *et al.*[7] in order to recover preservatives. A sample of 50–60 g was placed in a flask with 200 g of magnesium sulphate and 25 ml of phosphoric acid. Water was added until the volume was about 325 ml. The mixture was steam distilled into a separating funnel containing 50 ml of sodium hydroxide solution, with the distillation rate so adjusted that 725–750 ml could be collected within 90 min. The authors emphasize that this step is critical. The distillate was rendered acid to litmus by adding concentrated hydrochloric acid, and was then extracted with one 100 ml and four 50 ml portions of chloroform and ether (2:1). The combined extracts were gently evaporated on a steam bath (not to dryness), then transferred to a 50 ml volumetric flask and diluted to the mark with the chloroform–ether mixture. The solution was then ready for chromatography.

4.1.2 Paper chromatographic separations

Biefer[8] obtained satisfactory separations using Schleicher and Schuell chromatographic paper 2043b and developing with water and acetic acid (4:1). (Propyl gallate, $R_F = 0.70$; BHA, $R_F = 0.77$; NDGA, $R_F = 0.60$.)

When Whatman No. 1 paper was used in conjunction with a developing solvent of chloroform and acetic acid (99:1), the propyl gallate R_F was 0.38.

Ter Heide[9] was able to separate 3-BHA from 2-BHA by using Schleicher and Schuell 2043b paper and developing with petroleum ether and formic acid (2:1). (3-BHA, $R_F = 0.56$; 2-BHA, $R_F = 0.46$.)

Mitchell[10] applied immobile–mobile phase chromatography for resolution.

Adsorbent: Whatman No. 1 paper
Immobile phase: 10% soybean oil in ethyl ether
Mobile phase: methanol–water (80:20)

R_F: BHA = 0.56; BHT = 0.09; NDGA = 0.75; propyl gallate = 0.75. Using 100% water as the mobile phase, NDGA gave an R_F value of 0.05 and propyl gallate an R_F value of 0.36.

Dehority[11] used Whatman No. 1 paper which had been acetylated, and developed with acetone and water (60:40). The R_F values were determined relative to the propyl gallate value which was designated as 1.00. (BHA = 0.65; BHT = 0.38; NDGA = 0.81.)

Kiermeier and Jatczynski[12] resolved benzoic acid and some of its derivatives by paper chromatography.

Paper: Schleicher and Schuell 2043b
Solvent: *n*-butanol–ammonium hydroxide–water (70:20:10)

R_F: *p*-hydroxybenzoic acid = 0.06; benzoic acid = 0.35; *p*-chlorobenzoic acid = 0.52; methyl *p*-hydroxybenzoate = 0.65; ethyl *p*-hydroxybenzoate = 0.75; propyl *p*-hydroxybenzoate = 0.81; butyl *p*-hydroxybenzoate = 0.85.

The antioxidants can be detected on chromatograms by a variety of reagents. One can take advantage of their reducing properties by reacting them with solutions containing compounds such as ferro-ferricyanide, phosphomolybdic acid and tetrazolium salts. Reagents such as diazonium salts, aromatic aldehydes and Gibb's reagent, can also be used for their coupling activity with phenols which results in colored compounds. The benzoic acid compounds can be detected by spraying with 0.1 % solution of Rhodamine CG in ethanol, a solution of diazotized sulfanilic acid or a solution of bromphenol blue and methyl red.

Bighi and Saglietto[13] separated thiourea and ethylene thiourea by paper chromatography; there is some concern that ethylene thiourea (ETU) may prove to be a thyroid carcinogen.

Adsorbent: Whatman No. 3MM paper

Solvent: isobutyl alcohol–isopropyl alcohol–formic acid–water (40:30:10:20)

They were detected with alkaline silver nitrate (ETU, $R_F = 0.72$; thiourea, $R_F = 0.55$.)

4.1.3 Thin layer separations

Scheidt and Conroy[4] used Silica Gel HR plates with a solvent system of chloroform–methanol–acetic acid (90:10:2). R_F values obtained were: NDGA = 0.26; propyl gallate = 0.32; BHT = 0.90; BHA = 0.74. Salo and Salminen[14] also used silica gel plates, in conjunction with the solvent system Shell Sol A, *n*-propanol, acetic acid and formic acid (45:6:3:6). (NDGA, $R_F = 0.16$; BHT, $R_F = 0.85$; BHA, $R_F = 0.48$; tocopherol, $R_F = 0.66$.) Davidek *et al.*[15] used silica gel with two different solvent systems, giving the following results:

Solvent system 1: chloroform–acetic acid (9:1)

NDGA, $R_F = 0.12$; BHT, $R_F = 0.87$; 3-BHA, $R_F = 0.69$.

Solvent system 2: chloroform–acetic acid (75:25)

R_F: NDGA = 0.43; BHT = 0.92; BHA = 0.83; methyl gallate = 0.34; ethyl gallate = 0.43; propyl gallate = 0.51; octyl gallate = 0.64.

Davidek and Pokorny[16] also used polyamide layers.

Solvent system A: methanol–acetone–water (6:1:3)

R_F: propyl gallate = 0.56; BHA = 0.67; NDGA = 0.27.

Solvent system B: methanol–acetone–water (6:0.5:3.5)

R_F: propyl gallate = 0.42; BHA = 0.56; NDGA = 0.15.

Meyer[17] used silica gel and polyamide for antioxidant separations.

Adsorbent: Silica Gel G

Solvent system: petroleum ether–benzene–acetic acid (40:40:20)

R_F: methyl gallate = 0.06; ethyl gallate = 0.09; propyl gallate = 0.11; butyl gallate = 0.16; octyl gallate = 0.23; dodecyl gallate = 0.32; BHA = 0.63; BHT = 0.94; NDGA = 0.09; ascorbyl palmitate = 0.08.

Adsorbent: polyamide

Solvent system: methanol–acetone–water (60:20:20)

R_F: methyl gallate = 0.68; ethyl gallate = 0.55; propyl gallate = 0.52; butyl gallate = 0.48; octyl gallate = 0.18; dodecyl gallate = 0.06; BHA = 0.30; BHT = 0.24; NDGA = 0.26; ascorbyl palmitate = 0.04.

Pinella *et al.*[7] used a mixture of Kieselguhr and silica gel to resolve some preservatives.

Adsorbent: Kieselguhr G and Silica Gel (1 + 1)

Solvent A: hexane–acetic acid (96:4)

R_F: benzoic acid = 0.80; sorbic acid = 0.72; salicylic acid = 0.53; *p*-hydroxybenzoic acid = 0.09; methyl *p*-hydroxybenzoate = 0.10; ethyl *p*-hydroxybenzoate = 0.13; propyl *p*-hydroxybenzoate = 0.17.

Solvent B: *tert*-amyl alcohol–*tert*-butanol–ammonia–acetone (40:40:10:10)

R_F: benzoic acid = 0.38; sorbic acid = 0.31; salicylic acid = 0.60; *p*-hydroxybenzoic acid = 0.21; methyl *p*-hydroxybenzoate = 0.79; ethyl *p*-hydroxybenzoate = 0.82; propyl *p*-hydroxybenzoate = 0.86.

Wodich *et al.*[18] used polyamide layers with several solvent systems.

Solvent 1: benzene–acetone–acetic acid (5:45:1)

R_F: benzoic acid = 0.74; sorbic acid = 0.83; salicylic acid = 0.09.

Solvent 2: benzene–acetone–acetic acid (60:3:1)

R_F: benzoic acid = 0.40; sorbic acid = 0.47; salicylic acid = 0.08.

Solvent 3: pentane–hexane–acetic acid (10:10:3)

R_F: benzoic acid = 0.83; sorbic acid = 0.89; salicylic acid = 0.26.

Although caviar is not a common food, Brunstadt[19] developed an interesting thin layer chromatographic procedure for determining the boric acid in it, which is used as a mould inhibitor. The caviar was weighed into a mortar (100 mg sample), and extracted in 20 ml of water by grinding with the pestle. The washings were transferred to a glass-stoppered centrifuge tube and centrifuged for 5 min. The supernatant liquid was decanted into a 100 ml volumetric flask, 10 ml of water was added to the residue left in the centrifuge tube, which was then shaken, centrifuged and the supernatant layer added to the volumetric flask. This step was repeated.

Two ml of the sample solution were placed in a platinum dish with about 6 drops of 10% sodium hydroxide, mixed and then evaporated to dryness. The residue was dissolved in about 1.5 ml of a 0.12% curcumin solution, and to the resulting solution was added 1.5 ml of acetic acid and sulfuric acid (1:1). The solution was thoroughly mixed and allowed to stand for 15 min, then transferred to a beaker, diluted to 50 ml, and mixed again. For analysis, 40 μl of solution were spotted on a TLC plate alongside a blank extract.

Adsorbent: Silica Gel G on microscope slides 1″ × 3″

Developing solvent: ethanol–benzene (1:10)

The plates were sprayed with ammonium hydroxide, to show up the boric acid as a blue color. The plates were then sprayed with a solution of acetic acid and sulfuric acid (1:1) and the spot changed to a red-purple color. This particular procedure demonstrates some of the highly specific uses of thin layer chromatography.

Prepackaged and precooked foods are widely sold in supermarkets. In addition many fruits and vegetables are washed and then packaged in plastic-wrapped containers. As a result, the possibility always exists that chemicals from the various plastics may have been transferred onto the food. Thin layer chromatography can be used to determine the presence of the special anti-oxidants and UV absorbers used in plastic wrapping materials.

Simpson and Currell[20] used silica gel to resolve many of these compounds. Table 4.1 gives their commercial names and chemical types where known.

TABLE 4.1
Chemical formulae of antioxidants and UV absorbers

Nonox TBC	4-Methyl-2,6-*tert*-butyl phenol
Nonox WSP	2,2′-Methylene-bis-1-methylcyclohexyl-4-methylphenol
Nonox WSL	2-α-Methylcyclohexyl-4,6-dimethylphenol
Nonox CI	*N,N′*-di-β-Naphthyl-*p*-phenylenediamine
Nonox DPPD	*N,N′*-Diphenyl-*p*-phenylenediamine
Nonox NS	Phenolaldehyde–amine reaction product
Nonox WSO	Phenolic type
Irganox 1076	Hindered phenol molecule
Eastman DOBP	4-Dodecyloxy-2-hydroxybenzophenone
Eastman OPS	*p*-Octylphenyl salicylate
Uvinol 400	2,4-Dihydroxybenzophenone
Salol	Phenyl salicylate
Tinuvin P	Substituted benzotriazole
2246	2,2′-Methylene-bis-4-methyl-6-*tert*-butylphenol

(i) Antioxidants:

Adsorbent: Merck Kieselgel GF_{254} pre-coated plates

Solvent system: benzene–ethyl acetate–acetone (100:5:2)

R_F: Nonox SP = 0.61; Nonox TBC = 0.98; Nonox WSP = 0.85; Nonox EX = 0.56; Nonox WSL = 0.82; BHT = 0.98; 2246 = 0.75; Nonox CI = 0.65; Nonox DPPD = 0.66; Nonox OD = 0.56; Nonox Za = 0.43, 0.38; Santoflex 75 = 0.15, 0.61, 0.74; Santoflex AW = 0.80, 0.51, 0.55; Santoflex R = 0.26, 0.53; DLTDP = 0.61; Nonox NS = 0.52; Superlite = 0.64, 0.93; Polygard = 0.52, 0.61; Nonox

HO = 0.46, 0.64, 0.60, 0.99; Nonox WSO = 0.58; Irganox 1076 = 0.57, 0.75.

(ii) UV Absorbers:

Solvent system: chloroform–hexane (2:1)

R_F: Eastman DOBP = 0.39; Uvinul 400 = 0.08; Permyl B100 = 0.08; Salol = 0.45; Tinuvin HE = 0.71; Tinuvin P = 0.59; Eastman OPS = 0.65; UV 318 = 0.14; Cyasorb = 0.12; Cyasorb 1084 = 0.08; Uvinul N35 = 0.16.

Both antioxidants and UV absorbers were rendered visible by viewing under filtered UV radiation and spraying with a solution of 2,6-dichloroquinoneimine. The plate was then treated with a base, to produce colors varying from yellow to red to blue, depending upon where the molecule was substituted.

4.2 PLASTICIZERS

Plasticizers are added to food packaging materials and there is a possibility they may diffuse into the food. They are generally esters with a variety of structures. Peerebom[21] was able to resolve many plasticizers on Silica Gel G adsorbent using three different solvent systems.

Adsorbent: Silica Gel G

Solvent system 1: 2,2,4-trimethylpentane–ethyl acetate (90:10)

R_F: triacetin = 0.18; ethylphthaly ethyl glycolate = 0.22; acetyltriethyl citrate = 0.26; triphenyl phosphate = 0.33; tricresyl phosphate = 0.42; butylphthalyl butylglycolate = 0.43; 2-ethylhexyl diphenyl phosphate = 0.46; diethyl phthalate = 0.51; acetyl tributyl citrate = 0.53; di-*n*-butyl phthalate = 0.74; diisobutyl adipate = 0.83; dibutyl sebacate = 1.00 (reference compound); dinonyl phthalate = 1.01; di-(2-ethylhexyl)-phthalate = 1.14; butyl stearate = 1.61; Paraflex G 62 (epoxidized natural glycerides) = 0.09, 0.15, 0.25, 0.33, 0.60, 0.68, 0.73, 0.80, 0.92.

Solvent system 2: benzene–ethyl acetate (95:5)

R_F: triacetin = 0.34; ethylphthalyl ethyl glycolate = 0.66; acetyl triethyl citrate = 0.51; triphenyl phosphate = 0.80; tricresyl phosphate = 0.86; butylphthalyl butyl glycolate = 0.90; 2-ethylhexyl diphenyl phosphate = 0.77; diethyl phthalate = 0.79; acetyl tributyl citrate = 0.85; di-*n*-butyl phthalate = 1.03; diisobutyl adipate = 0.86; dibutyl sebacate = 1.0 (reference compound); dinonyl phthalate = 1.18; di-(2-ethylhexyl)phthalate = 1.16; butyl stearate = 1.23; Paraflex G 62 = 0.06, 0.13, 0.20, 0.26, 0.40, 0.47, 0.65, 0.72, 1.04.

Solvent system 3: hexane–dibutyl ether (20:80)

R_F: triacetin = 0.17; ethylphthalyl ethyl glycolate = 0.30; acetyl triethyl citrate = 0.29; triphenyl phosphate = 0.50; tricresyl phosphate = 0.69; butylphthalyl butyl glycolate = 0.65; 2-ethylhexyl diphenyl phosphate = 0.58; diethyl phthalate = 0.60; acetyl tributyl citrate = 0.70; di-*n*-butyl phthalate = 0.84; diisobutyl adipate = 0.85; dibutyl sebacate = 1.0 (reference compound); dinonyl phthalate = 1.14; di-(2-ethylhexyl)phthalate = 1.15; butyl stearate = 1.28; Paraflex G 62 = 0.06, 0.14, 0.28, 0.42.

The plasticizers were extracted from the polymer with ethyl ether or benzene. The plates were impregnated with a water-soluble indicator called Ultraphor which fluoresced, when irradiated with UV radiation at 365 nm, against which the plasticizers could be seen as quenched or nonfluorescent spots.

Braun[22] also used Silica Gel G but used a single component solvent of methylene chloride instead.

R_F: dimethyl phthalate = 0.51; dibutyl phthalate = 0.69; dihexyl phthalate = 0.80; dioctyl phthalate = 0.86; di-(2-ethylhexyl)phthalate = 0.85; didecyl phthalate = 0.85; diisodecyl phthalate = 0.84; trioctyl phosphate = 0.23; dipheny octyl phosphate = 0.42; triphenyl phosphate = 0.47; diphenyl cresyl phosphate = 0.51; tricresyl phosphate = 0.53; dioctyl adipate = 0.42; di-(2-ethylhexyl)-adipate = 0.44; dinonyl adipate = 0.44; adipic acid polyester = 0.02; dibutyl sebacate = 0.41; dioctyl sebacate = 0.61; di-(2-ethylhexyl)sebacate = 0.61; sebacic acid polyester = 0.02; triethyl citrate = 0.12; tributyl citrate = 0.14; triethyl acctyl citrate = 0.15; tributyl acetyl citrate = 0.24; tri-(2-ethylhexyl)acetyl citrate = 0.46; glycerol triacetate = 0.13.

4.3 SWEETENERS

Artificial sweeteners have long been used in diets for diabetics, and more recently, in ordinary foods and beverages to reduce their calorie content. Recently evidence has been obtained to indicate that some of these sweeteners can be carcinogenic when fed in large amounts to rats.

4.3.1 Isolation

The sweetening solutions themselves can be analyzed by evaporating any alcohol that may be present and redissolving them in a solvent more suitable for spotting. Beverages were analyzed[23] by placing 50 ml of decarbonated sample into a 125 ml separating funnel, and adding 10 ml of sulfuric acid and water (1:1). The solution was cooled and then extracted with two 50 ml volumes of petroleum ether. The ether layer was drawn off and discarded. The aqueous layer was neutralized by adding 5 ml of a 50% sodium hydroxide solution and then extracted with two 50 ml portions of ethyl acetate, which were then filtered and concentrated for analysis.

The sweeteners may also be isolated by distillation,[24] in the following manner: 300 ml of a drink or other consumable liquid are placed in a 1 liter boiling flask with antibumping chips, then enough 10N sodium hydroxide is added to raise the pH to 14. The sweeteners are distilled into a vessel containing 2 ml of 6N hydrochloric acid. Approximately 35 ml of distillate are collected in about 20 min. The volume is reduced on a hot plate at just below boiling point until 3–5 ml are left. The concentrate is transferred to a separating funnel with four 25 ml volumes of water. Sodium hydroxide solution is added to raise the pH to 14, and the solution is then extracted with 20 ml of methylene chloride and filtered through sodium sulfate prewetted with solvent. It is further extracted twice with 10 ml volumes of methylene chloride, and filtered through sodium sulfate. After each filtration the sodium sulfate must be rinsed with 10 ml of methylene chloride. The total filtrates are collected in a concentrator, such as a Kuderna-Danish, and concentrated to the desired volume in readiness for chromatography.

4.3.2 Paper chromatography

Mitchell[25] used Whatman No. 1 paper in conjunction with the solvent mixture acetone, ethyl acetate and ammonia (8:1:1) for separating three artificial sweeteners. (Saccharin, $R_F = 0.33–0.38$; Dulcin, $R_F = 0.89–0.94$; cyclamate, $R_F = 0.18–0.22$.) They were made visible by placing the chromatograms in a tank containing bromine fumes, evaporating off the excess bromine in a hood, spraying with alcoholic fluorescein and then observing the spots under UV radiation at 365 nm.

4.3.3 Thin layer chromatography

Nagasawa *et al.*[26] separated the sweeteners with various types of adsorbent layers.

Adsorbent: microcrystalline cellulose

Solvent system A: ethyl acetate–ammonia–acetone (1:1:8)

Saccharin, $R_F = 0.31$; Dulcin, $R_F = 0.82$; cyclamate, $R_F = 0.15$.

Solvent system B: dioxane–pyridine–water (7:2:1)

Saccharin, $R_F = 0.39$; Dulcin, $R_F = 0.80$; cyclamate, $R_F = 0.24$.

Solvent system C: pyridine–ethanol–water (6:3:1)

Saccharin, $R_F = 0.64$; Dulcin, $R_F = 0.79$; cyclamate, $R_F = 0.47$.

A polyamide adsorbent layer with a solvent system of benzene, ethyl acetate and formic acid (5:10:2) gave an R_F of 0.69 for saccharin, 0.81 for Dulcin and 0.41 for cyclamate. Silica gel in conjunction with the solvent mixture of acetone and 10% ammonium hydroxide (9:1) gave an R_F of 0.59 for saccharin, 0.84 for Dulcin and 0.24 for cyclamate. Alumina was also successfully used for resolving the three sweeteners.

Solvent system: isopropanol–ammonium hydroxide (4:1)

Saccharin, $R_F = 0.30$; Dulcin, $R_F = 0.70$; cyclamate, $R_F = 0.14$.

DEAE-cellulose used with 0.5M ammonium formate solution at pH 6.5, gave the following R_F values: saccharin = 0.36; Dulcin = 0.53; cyclamate = 0.74.

The sweeteners can be visualized on a thin layer plate by observing the chromatograph under UV radiation at 253.7 nm. Saccharin gives a blue fluorescent spot and Dulcin a quenched spot. If the plate is sprayed with a 0.1% ethanolic solution of Pinacryptol Yellow and then observed with 365 nm radiation, saccharin appears as an orange spot, cyclamate as an orange fluorescent spot on a light greenish background, and Dulcin as a violet quenched spot.

Korbelak and Bartlett[27] used silica gel as the adsorbent layer and developed with *n*-butanol, 95% ethanol, 28% ammonium hydroxide and water (40:4:1:9). They obtained an R_F value of 0.41 for saccharin, 0.75 for Dulcin and 0.30 for cyclamate; they were also able to get an R_F value of 0.84 for the P 4000 sweeteners (5-nitro-2-*n*-propoxyaniline).

4.4 ANTIBIOTICS

Antibiotics are sometimes used for reducing the spoilage of meat, poultry and fish. Two of the most widely used are oxytetracycline and chlortetracycline and, because these compounds are also used for the treatment of infections in humans, many toxicologists are wary of allowing the human consumption of their residues in food. A number of countries still allow antibiotic residue values of 5–10 ppm in meat, poultry and fish. In the United States, however, these residues are altogether disallowed in food.

4.4.1 Paper chromatography

This technique can be used for resolving the two antibiotics mentioned above. Seltzer and Wright[28] used Whatman No. 1 paper with a developing solvent of chloroform, nitromethane and pyridine (10:20:30). Before developing, the paper was wetted with pH 4.5 McIlvaine's buffer and then blotted dry. R_F values obtained were: chlortetracycline = 0.50; oxytetracycline = 0.13. The spots were detected by their fluorescence after exposing the paper to ammonia vapor.

Bohonos *et al.*[29] used Eaton-Dikeman No. 613 paper soaked in a 0.3M pH 3 phosphate buffer and air-dried. The chromatogram was developed in *n*-butanol saturated with water. (Oxytetracycline, $R_F = 0.40$; chlortetracycline, $R_F = 0.65$.) The compounds were detected by bioautography using *Bacillus cereus* on pH 6.0 nutrient agar.

4.4.2 Thin layer chromatography

This technique was used by Gyanchandani *et al.*[30] for the separation of the antibiotics.

Adsorbent: Kieselguhr G (50 g mixed with 0.1M aqueous EDTA and 20% PEG 400 in glycerin (95:5) (dried at room temperature after coating).

Solvent system: methylene chloride–ethyl formate–ethanol (9:9:2) (saturated with pH 4.7 McIlvaine's buffer).

Chlortetracycline, $R_F = 0.60$; oxytetracycline, $R_F = 0.20$.

The compounds can be detected with UV illumination or by spraying, first with Fast Blue B (diazonium salt) and then with 0.1N aqueous sodium hydroxide, with which they give a pink color.

4.5 NITROSAMINES

Nitrates are added to cured meat because they can be converted to nitrites by bacterial or enzymatic reduction. Nitrites in turn give a particular color and taste to the meat and are anticlostridial. This treatment of meat has been used for many years, but with the incidence of cancer showing a statistical up-swing, toxicologists are concerned about the possibility of nitrosamine formation in the presence of nitrite anions under acid conditions, nitrosamines having been linked with cancer formation.

Alcoholic beverages brewed or distilled in certain localities have also been found to contain nitrosamines. Sen and Dalpe[31] developed a method for determining dimethyl-, diethyl- and dipropyl- nitrosamines. A 100 ml aliquot of the alcoholic drink to be analyzed was placed in a 1 liter distillation flask and 12 g of sodium hydroxide was dissolved in the sample. The mixture was distilled and about 50 ml of distillate collected. Five grams of potassium carbonate were dissolved in the distillate and extracted with two 50 ml volumes of dichloromethane. The combined extracts were shaken with 40 ml of glycine–hydrochloric acid buffer and the aqueous layer discarded. The organic layer was washed with 40 ml of 20% potassium carbonate solution and the final dichloromethane layer dried over anhydrous sodium sulfate. The dry extract was filtered through Whatman No. 1 filter paper and collected in a 250 ml distillation flask. A Snyder column was fitted to the flask and the volume evaporated down to a few ml in a hot water bath at 55–60 °C. The concentrate was transferred to a small pear-shaped container fitted with a micro-Snyder column and carefully evaporated further to a few drops. The sides of the container were washed with dichloromethane till the volume of its contents increased to 0.5 ml. The solution was then ready for chromatography.

Further clean-up or isolation could be carried out with a basic alumina column (Woelm activity I), prepared by packing a glass column of 1–1.2 cm diameter with a 25–30 mm layer of the basic alumina. The solution resulting from Sen and Dalpe's distillation was diluted to 10 ml, transferred to the column and allowed to flow completely through. The column was then eluted with three different combinations of pentane and dichloromethane (41.5 ml + 2.5 ml, 45 ml + 5 ml, and 40 ml + 10 ml of pentane and dichloromethane, respectively) at a flow rate of 2–3 ml min^{-1}. The nitrosamines were eluted from the column by passing through 75 ml of dichloromethane at a rate of 2–3 ml min^{-1}. The solution was evaporated down to 0.3 ml and was ready for thin layer chromatography.

Adsorbent: MN Silica Gel G-HR

Solvent system: hexane–diethyl ether–dichloromethane (40:30:20)

R_F: dimethylnitrosamine = 0.28; diethylnitrosamine = 0.48; dipropylnitrosamine = 0.68.

The spots were made visible by spraying with Griess reagent and irradiating with UV radiation for 10 min. Nitrosamines appeared as purple spots.

Preussmann *et al.*[32] developed a separation method for many types of nitrosoamines.

Adsorbent: Silica Gel G

Solvent system 1: hexane–diethyl ether–methylene chloride (4:3:2)

R_F: nitroso-dimethylamine = 0.24; nitrosodiethylamine = 0.49; nitroso-di-*n*-propylamine = 0.69; nitroso-di-isopropylamine = 0.64; nitroso-di-*n*-butylamine = 0.77; nitroso-diamylamine = 0.82; nitroso-methylethylamine = 0.36; nitroso-methylpropylamine = 0.46; nitroso-methylbutylamine = 0.51; nitrosomethylamylamine = 0.57; nitroso-methylheptylamine = 0.60; nitroso-methylbenzylamine = 0.53.

Solvent system 2: hexane–diethyl ether–methylene chloride (5:7:10)

R_F: *N*-nitroso-*N'*-methylpiperazine = 0.04; di-nitrosopiperazine = 0.27; *N*-nitroso-*N'*-carbethoxypiperazine = 0.35; nitrosomorpholine = 0.40; nitroso-pyrrolidine = 0.41; nitroso-proline-ethylester = 0.52; nitroso-piperidine = 0.63; nitroso-indoline = 0.74.

Solvent system 3: hexane–diethyl ether–methylene chloride (10:3:2)

R_F: *p*-nitrosomethyl aminobenzoic acid = 0; *p*-nitrosomethyl aminobenzaldehyde = 0.22; nitroso-methyl-2-phenylethylamine = 0.29; nitroso-methylphenylamine = 0.63; nitroso-ethylphenylamine = 0.72; nitroso-diphenylamine = 0.80.

Two chromogenic reagents were used for detection. The developed plate was illuminated with high intensity UV radiation, and then sprayed with a mixture of 1.5% ethanolic diphenylamine and a 0.1% solution of palladium chloride in 0.2% sodium chloride. Blue to violet spots were formed. The Greiss reagent consisted of 1% sulfanilic acid in 30% acetic acid mixed with 0.1% naphthylamine in 30% acetic acid. After irradiating with UV radiation the plate was sprayed with the mixture and the nitrosoamines became visible as red to violet spots.

4.6 FOOD DYES

In 1900 around eighty dyes were used in the United States for coloring food products. When the Food and Drug Act of 1906 was made legally binding only seven of these dyes were still permitted for use in foods. These dyes were declared acceptable only after toxicology studies had been performed with the methods available at the time. The seven original dyes were Orange I, Erythrosine, Ponceau 3R, Amaranth, Indigotine, Naphthol Yellow and Light Green. Later, in 1938 the new Food, Drug and Cosmetic Act became law and formally created three classes of usable dyes. Those used for foods, drugs and cosmetics were designated FD and C colors; those intended for drugs and cosmetics D and C colors; while external drug and cosmetic dyes were called Ext. D and C colors.

Today, artificial colorings are used in many foods particularly in deserts and beverages. Indeed, beverages are available which contain artificial flavorings, colorings and sweeteners. Dyes that are used as food additives are mostly water-soluble and consist of four molecular families: azo dyes such as FD and C Red No. 2 (Amaranth), triphenylmethane dyes such as FD and C Blue No. 1 (Brilliant Blue), fluorescein type dyes, such as FD and C Red No. 3 (Erythrosine) and sulfonated indigo dyes such as FD and C Blue No. 2 (Indigotine).

At present, all food dyes are undergoing renewed scrutiny because of their possible carcinogenic effects and because there are indications that some red dyes produce hyperactivity in children.

Figure 4.3 shows the representative structures of the food dyes and Table 4.2

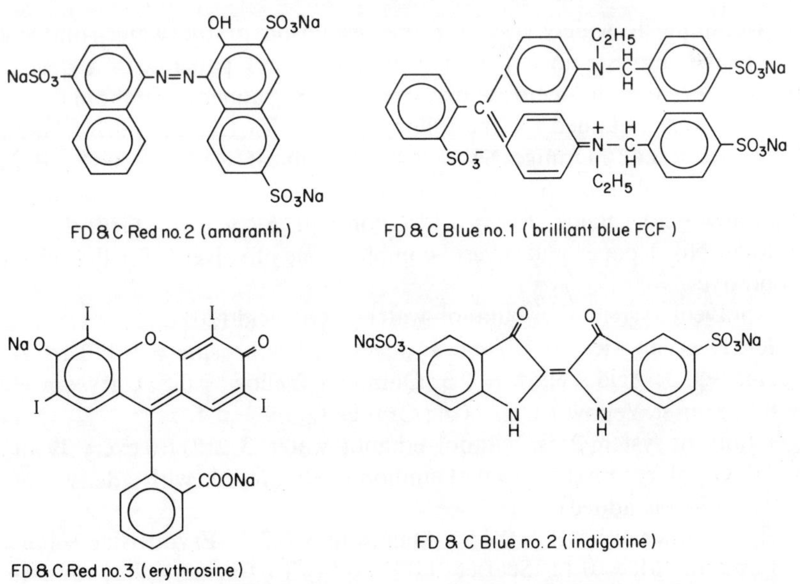

Fig. 4.3. Structural formulae of some food dyes.

gives the FD and C designations of colors approved by the European Economic Community. It can be seen from the preceding account that the use of food dyes could affect the suitability of a food product for import or export.

TABLE 4.2
Food dyes approved by the European Economic Community

Type of Dye	Color Index No.
FD & C Red No. 2	16185
FD & C Red No. 3	45430
FD & C Yellow No. 6	15985
FD & C Yellow No. 5	19140
FD & C Blue No. 2	73015
Carmoisine (Azorubine)	14720
Ponceau 4R (Cochineal Red A)	16255
Scarlet GN	14815
Ponceau 6R	16290
Orange GGN	15980
Fast Yellow (Acid Yellow G)	13015
Chrysoine S (Resorcine Yellow)	14270
Quinoline Yellow WS (D & C Yellow 10)	47005
Patent Blue V	42051
Indanthrene Blue RS	69800
Brilliant Black BN	28440
Black 7984	—

4.6.1 Paper chromatography

This technique is widely used for the resolution of the water-soluble food dyes. Thaler and Sommer[33] used Whatman No. 1 paper with a developing solvent of 2% trisodium citrate dissolved in a 5% aqueous ammonia solution. R_F values were obtained for the following dyes: Tartrazine = 0.70; Naphthol Yellow SK = 0.42; Orange SY = 0.48; Cochineal Red A = 0.58; Ponceau SX = 0.28; Azorubin S = 0.22; Fast Red E = 0.30; Amaranth = 0.46; Indigotin = 0.20; Black PN = 0.12; Ponceau 6R = 0.66. Griffiths[34] used Whatman No. 1 paper and several combinations of solvents for the resolution of food dyes.

Solvent system 1: *n*-butanol–water–acetic acid (20:12:5)

R_F: Ponceau 4R = 0.4; Amaranth = 0.3; Erythrosine = 1.0; Tartrazine = 0.3; Acid Yellow = 0.3; Quinoline Yellow = 0.5; Chrysoin = 0.5; Sunset Yellow FCF = 0.4; Orange GGN = 0.4.

Solvent system 2: isobutanol–ethanol–water (3:2:2) (to every 99 ml of solvent mixture 1 ml of ammonium hydroxide with a density of 0.880 is added)

R_F: Ponceau 4R = 0.4; Amaranth = 0.2; Erythrosine = 0.6; Ponceau 6R = 0.1; Scarlet GN = 0.5; Tartrazine = 0.2; Sunset Yellow FCF = 0.5; Patent Blue = 0.6.

Solvent system 3: methyl ethyl ketone–acetone–water–ammonium hydroxide (70:30:30:0.2)

R_F: Ponceau 4R = 0.3; Amaranth = 0.1; Erythrosine = 0.6; Tartrazine = 0.1; Acid Yellow = 0.4; Quinoline Yellow = 0.6; Chrysoin = 0.3; Sunset Yellow FCF = 0.4; Orange GGN = 0.4; Patent Blue = 0.6.

Solvent system 4: methyl ethyl ketone–acetone–water (70:30:30)

R_F: Ponceau 4R = 0.3; Amaranth = 0.2; Erythrosine = 1.0; Ponceau 6R = 0.1; Scarlet GN = 0.6; Tartrazine = 0.2; Sunset Yellow FCF = 0.5; Patent Blue = 0.7.

McNeal, in order to extract dyes from meat products[35] for separation by paper chromatography, slurried a meat sample with a minimum amount of water or 80% ethanol and allowed it to stand for 5 min. The slurry was filtered and the filtrate divided into three equal portions, which were separately evaporated just to dryness on a steam bath and the residue in each case taken up with ethanol to be spotted on chromatography paper.

Adsorbent: Whatman No. 1 paper

Solvent A: ammonium hydroxide–2.5% sodium citrate–water (45:10:45)

R_F: FD and C Red 1 = 0.15; FD and C Red 2 = 0.55; FD and C Red 3 = 0.05; FD and C Red 4 = 0.42; FD and C Red 40 = 0.35; FD and C Yellow 1 = 0.70; FD and C Yellow 4 = 0.20; FD and C Yellow 5 = 0.85; FD and C Yellow 6 = 0.77; FD and C Blue 1 = 0.95; FD and C Blue 2 = 0.18; FD and C Green 2 = 1.0; FD and C Green 3 = 1.0; FD and C Violet 1 = 0.80; methyl violet = 0.03; Orange B = 0.57; Orange 1 = 0.36; Orange 2 = 0.36.

Solvent B: *n*-propanol–ethyl acetate–water (6:1:3)

R_F: FD and C Red 1 = 0.32; FD and C Red 2 = 0.20; FD and C Red 3 = 0.70; FD and C Red 4 = 0.50; FD and C Red 40 = 0.45; FD and C Yellow 1 = 0.50; FD and C Yellow 2 = 0.96; FD and C Yellow 5 = 0.21; FD and C Yellow 6 = 0.35; FD and C Blue 1 = 0.46; FD and C Blue 2 = 0.21; FD and C Green 2 = 0.39; FD and C Green 3 = 0.46; FD and C Violet 1 = 0.65; methyl violet = 1; Orange B = 0.45; Orange 1 = 0.61; Orange 2 = 0.64.

Alcoholic beverages may not be considered foods but they are widely consumed all over the world and dyes are used to color liquors and wines. An interesting technique to assay such colorings was applied by Tewari *et al.*[36] A 10 ml sample was evaporated just to dryness and the residue dissolved in 0.5 ml of 50% ethanol. An aliquot of this ethanolic solution was spotted on Whatman No. 1 chromatographic paper along with standards of the allowable dyes. The ascending technique was used with a solvent system of *n*-butanol, acetic acid and water (4:1:5). R_F values obtained showed that most of the dyes used for coloring alcoholic beverages are a mixture of different colors. For example, 'Bush' Dark-Brown (a commercial product) separated into a range of colors with differing R_F values, viz, yellow = 0.04, violet = 0.0, orange = 0.26, deep-pink = 0.38 and greenish-blue = 0.70; 'Bush' Permical (an orange-red powder) separated into four colors with the following R_F values: red = 0.03; orange = 0.26; deep-pink = 0.38; red = 0.77.

4.6.2 Thin layer chromatography

A similar study on beverages[36] was carried out using Silica Gel G as the adsorbent and a developing solvent of *n*-butanol, acetic acid and water (4:1:5). The 'Bush' Dark-brown product separated into a slightly different range of colours compared to the paper chromatographic assay: lemon-yellow, $R_F = 0.08$; violet, $R_F = 0.17$; orange, $R_F = 0.27$; green, $R_F = 0.32$; violet-pink, $R_F = 0.38$; red, $R_F = 0.73$. The 'Bush' Permical dye separated into the following component colors: red, $R_F = 0.06$; orange, $R_F = 0.27$; violet pink, $R_F = 0.38$; red, $R_F = 0.73$.

Cellulose thin layers were used by Peniver[37] for the resolution of water-soluble dyes.

Adsorbent: cellulose

Solvent system 1: ethyl acetate–*n*-butanol–pyridine–water (5:5:6:5)

R_F: FD and C Blue 1 = 0.53; FD and C Blue 2 = 0.33; FD and C Green 3 = 0.49; FD and C Red 2 = 0.28; FD and C Red 3 = 0.94; FD and C Yellow 5 = 0.26; FD and C Yellow 6 = 0.47; FD and C Violet 1 = 0.61.

Solvent system 2: ethyl acetate–*n*-propanol–ammonium hydroxide ($d = 0.880$)–water (35:35:20:20)

R_F: FD and C Blue 1 = 0.72; FD and C Blue 2 = 0.47; FD and C Green 3 = 0.57; FD and C Red 2 = 0.28; FD and C Red 3 = 0.79; FD and C Yellow 5 = 0.15; FD and C Yellow 6 = 0.62; FD and C Violet 1 = 0.80.

Solvent system 3: *n*-propanol–ethyl acetate–ammonium hydroxide ($d = 0.880$) (65:75:60)

R_F: FD and C Blue 1 = 0.56; FD and C Green 3 = 0.38; FD and C Red 2 = 0.10; FD and C Red 3 = 0.68; FD and C Yellow 6 = 0.41; FD and C Violet 1 = 0.66.

Wollenweber[38] used Cellulose MN 300 layers with different solvent systems.

Solvent A: 2.5% trisodium citrate (aqueous solution)–25% ammonium hydroxide (4:1)

R_F: Fast Yellow = 0.57; Tartrazine = 0.76; Quinoline Yellow = 0.13; Chrysoin S = 0.40; Yellow 27175 N = 0.34; Orange GGN = 0.52; Sunset Yellow FCF = 0.48; Azorubin S = 0.16; Fast Red E = 0.24; Amaranth S = 0.39; Cochineal Red A = 0.62; Ponceau 6R = 0.77; Scharlach GN = 0.85; Indigotin = 0.12; Brilliant Black BN = 0.15.

Solvent B: *n*-propanol–ethyl acetate–water (6:1:3)

R_F: Fast Yellow = 0.53; Tartrazine = 0.26; Quinoline Yellow =

0.35; Chrysoin S = 0.78; Yellow 27175 N = 0.33; Orange GGN = 0.55; Sunset Yellow FCF = 0.54; Azorubin S = 0.64; Fast Red E = 0.62; Amaranth S = 0.30; Cochineal Red = 0.35; Ponceau 6R = 0.13; Scharlach GN = 0.64; Indigotin = 0.33; Brilliant Black BN = 0.20.

Martin and Figert[39] detected dyes in distilled spirits by chromatography on precoated cellulose plates (EM Laboratories). A 50 ml sample of a flavoring extract or a 100 ml sample of an alcoholic beverage was placed in a 250 ml beaker. Hydrochloric acid was added till the pH dropped to 2. A 12 inch piece of white wool yarn was then placed in the beaker, the solution brought to boil and kept boiling until its volume was reduced to 25 ml. The piece of wool yarn was then taken from the beaker, rinsed thoroughly with cold water and examined for adhering dye. If after rinsing the wool remained white, it was assumed that no dyes of so-called coal-tar origin were present. If, however, the wool remained colored, the adhering dye was removed by washing out with 25 ml of 10% ammonium hydroxide. This was done by allowing the wool to stand in the basic solution for 15 min. The dye solution in the beaker was evaporated down to about 2 ml, and the sample was then ready for spotting.

Two solvent systems were used.

Solvent system 1: ethyl acetate, *n*-butanol, pyridine and water (25:25:30:25)

R_F values: FD and C Blue 1 = 0.42; FD and C Blue 2 = 0.25; FD and C Red 2 = 0.21; FD and C Red 3 = 0.93; FD and C Red 4 = 0.50; FD and C Green 3 = 0.46; FD and C Yellow 5 = 0.13; FD and C Yellow 6 = 0.42; FD and C Violet 1 = 0.67.

Solvent system 2: *n*-propanol, triethylamine and water (62:10:28).

R_F values: FD and C Blue 1 = 0.80; FD and C Blue 2 = 0.36; FD and C Red 2 = 0.43; FD and C Red 3 = 1; FD and C Red 4 = 0.54; FD and C Green 3 = 0.68; FD and C Yellow 5 = 0.32; FD and C Yellow 6 = 0.74; FD and C Violet 1 = 0.96.

Graham and Nya[40] resolved the dyes on pre-coated Merck silica gel plates with a solvent system consisting of *n*-butanol, ethyl methyl ketone, ammonium hydroxide ($d = 0.880$) and water (5:3:1:1), using either the sandwich tank technique or the unsaturated tank technique to develop their plates. Sandwich tank development gave these R_F values: Black PN = 0.04; Violet BNP = 0.21; Blue VRS = 0.19; Green S = 0.06; Yellow 2G = 0.08; Naphthol Yellow = 0.23; Tartrazine = 0.05; Yellow RFS = 0.19; Sunset Yellow = 0.18; Orange G = 0.18; Orange RN = 0.35; Erythrosine = 0.35; Ponceau MX = 0.36; Red 2G = 0.16; Fast Red E = 0.23. R_F values for unsaturated tank development were as follows: Black Pn = 0.11; Violet BNP = 0.30; Blue VRS = 0.31; Green S = 0.13; Yellow 2G = 0.14; Naphthol Yellow = 0.35; Tartrazine = 0.11; Yellow RFS = 0.27; Sunset Yellow = 0.29; Orange G = 0.26; Orange RN = 0.68; Erythrosine = 0.69; Ponceau MX = 0.68; Red 2G = 0.25; Fast Red E = 0.36.

Pietsch and Meyer[41] used Silica Gel G in combination with a developing solvent of ethyl acetate, methanol and 4.6N ammonium hydroxide (50:15:10). R_F values were obtained for the following dyes: Fast Yellow = 0.48; Tartrazine = 0.16; Quinoline Yellow = 0.44; Chrysoin S = 0.50; Orange GGN = 0.55; Sunset Yellow FCF = 0.48; Azorubin S = 0.49; Fast Red E = 0.59; Amaranth S = 0.17; Cochineal Red A = 0.25; Ponceau 6R = 0.05; Scharlach GN = 0.62; Indigotin = 0.43; Brilliant Black BN = 0.10. The dyes were of course detectable by their colors but they were also observed under UV illumination, in which a number of them changed color or fluoresced. Treatment of the spots with acids or bases caused other color changes which further aided positive identification.

Takeshita *et al.*[42] developed a method for the isolation and detection of water-soluble acid dyes in foods. DEAE-Sephadex A-25 was allowed to absorb water, then equilibrated with 0.5N ammonium hydroxide and washed with more water until no chloride ion was detectable by the 1% silver nitrate test. The Sephadex was then equilibrated with 1N acetic acid overnight and again washed with water. A glass column of diameter 0.4 cm and 3.5 cm long was filled with a mixture which was one part Sephadex to one part distilled water. A 5–10 g sample was ground with a pestle and mortar and then filtered through 300 mesh cloth. (The sample could also be mechanically homogenized before filtering.) If it contained much suspended fat this could be removed by partitioning from the aqueous solution into several volumes of hexane. The prepared filtrate was slowly added to a 50 ml volumetric flask containing 1 ml of the Sephadex suspension, until it was no longer adsorbed by the Sephadex and the color was visible in the supernatant. The mixture was centrifuged, the supernatant discarded, and the Sephadex washed with 30 ml of water which was also discarded. This treatment was repeated until no turbidity was evident in the washings. To the Sephadex in the centrifuge tube was added 10 ml of 50% aqueous isopropanol and the slurry transferred to the small column. The column was eluted with 2N hydrochloric acid and isopropanol (1:1) at a flow rate of about 0.1–0.15 ml min^{-1}. As soon as the dyes began to wash through. a 0.5 ml volume of the eluate was collected and neutralized with 0.25 ml volume of 4N ammonium hydroxide and isopropanol (1:). A 0.5 μl volume of the neutralized solution was used for spotting on the thin layer plate.

Precoated Polyamide UA plates were used with a developing solvent of pyridine, methanol, 28% ammonium hydroxide and water (5:6:1:16). The authors gave no absolute R_F values but offered graph comparisons of several chromatograms of the same dyes run with different ratios of pyridine in the developing solvent. They claimed that the spots showed no tailing in the above solvent system.

The following water-soluble acid dyes were chromatographed: Erythrosine(CI 45430); eosine(CI 45380); phloxine(CI 45410); Rose Bengal(CI 45440); Acid Red(CI 45100); Guinea Green B(CI 42085); Light Green SF(CI 42095); Fast Green FCF(CI 42053); Brilliant Blue FCF(CI 42090); Acid Violet

6B(CI 42640); Ponceau 3R(CI 16155); Ponceau R(CI 16150); Ponceau SX(CI 14700); amaranth(CI 16185); New Coccine(CI 16255); tartrazine(CI 19140); Sunset Yellow FCF(CI 15985); Orange I(CI 14600); Napthol Yellow S(CI 10316); Indigo Carmine(CI 73015).

Only a few of the various groups of food additives currently in use have been considered in this chapter, and these have been chosen because they most concern toxicologists who may be studying the role of food additives as carcinogens. However, it is probable that in the near future many other additives such as flavorings, texturizing agents and nutrients will be investigated for toxic effects and no doubt both paper and thin layer chromatography will prove useful in supplementing the toxicological data on them.

REFERENCES

1. R. Ter Heide, *Fette Seifen Anstrichm.* **60**, 360 (1958).
2. A. Seher, *Fette Seifen Anstrichm.* **65**, 1002 (1963).
3. M. A. Philips and R. D. Hinkel, *J. Agric. Food Chem.* **5**, 379 (1957).
4. S. A. Scheidt and H. W. Conroy, *J. Assoc. Off. Anal. Chem.* **49**, 807 (1966).
5. W. G. Schwien, B. J. Miller and H. W. Conroy, *J. Assoc. Off. Anal. Chem.* **49**, 809 (1966).
6. D. F. McCauley, T. Fazio, J. W. Howard, F. M. DiCiurcio and J. Ives, *J. Assoc. Off. Anal. Chem.* **50**, 243 (1967).
7. S. J. Pinella, A. D. Falco and G. Schwartzman, *J. Assoc. Off. Anal. Chem.* **49**, 829 (1966).
8. K. W. Biefer, *Mitt. Geb. Lebensmittelunters. Hyg.* **53**, 243 (1962).
9. R. Ter Heide, *Seifen Oele Fette Wachse* **92**, 890 (1966).
10. L. C. Mitchell, *J. Assoc. Off. Agric. Chem.* **40**, 908 (1957).
11. B. A. Dehority, *J. Chromatogr.* **2**, 384 (1959).
12. F. Kiermeier and R. Jatczynski, *Z. Lebensm. Unters. Forsch.* **113**, 330 (1960).
13. C. Bighi and G. Saglietto, *J. Chromatogr.* **17**, 18 (1965).
14. T. Salo and K. Salminen, *Z. Lebensm. Unters. Forsch.* **125**, 167 (1964).
15. J. Davidek, G. Janicek and E. Davidkova, *Z. Lebensm. Unters. Forsch.* **131**, 345 (1967).
16. J. Davidek and J. Pokorny, *Z. Lebensm. Unters. Forsch.* **115**, 113 (1961).
17. H. Meyer, *Dtsch. Lebensm. Rundsch.* **57**, 170 (1961).
18. H. Wodich, E. Gnauer and E. Galinovski, *Z. Lebensm. Unters. Forsch.* **133**, 317 (1967).
19. J. W. Brunstad, *J. Assoc. Off. Anal. Chem.* **51**, 987 (1968).
20. D. Simpson and B. R. Currell, *Analyst* **96**, 515 (1971).
21. J. W. C. Peerebom, *J. Chromatogr.* **4**, 323 (1960); **11**, 55 (1963).
22. T. Braun, *Chimia* **19**, 77 (1965).
23. *Official Methods of Analysis of the Association of Official Analytical Chemists*, 12th Edn, Section 20.154, p. 376, 1975.
24. *Official Methods of Analysis of the Association of Official Analytical Chemists*, 12th Edn, Section 20.165, p. 378, 1975.
25. L. C. Mitchell, *J. Assoc. Off. Agric. Chem.* **38**, 943 (1955).
26. K. Nagasawa, H. Yoshidome and K. Anryu, *J. Chromatogr.* **52**, 173 (1970).
27. T. Korbelak and J. N. Bartlett, *J. Chromatogr.* **41**, 124 (1969).
28. G. B. Seltzer and W. W. Wright, *Antibiot. Chemother.* **7**, 292 (1957).

29. N. Bohonos, A. C. Dornbush, L. I. Feldman, J. H. Martin, E. Pelcak and J. H. Williams, *Antibiot. Ann.* 49–55 (1953–1954).
30. N. D. Gyanchandani, I. J. McGilveray and D. W. Hughes, *J. Pharm. Sci.* **59**, 224 (1970).
31. N. P. Sen and C. Dalpe, *Analyst* **97**, 216 (1972).
32. R. Preussmann, G. Neurath, G. Wulf-Lorentzen, D. Daiber and H. Hengy, *Z. Anal. Chem.* **202**, 187 (1964).
33. H. Thaler and G. Sommer, *Z. Lebensm. Unters. Forsch.* **97**, 345, 441 (1953).
34. M. H. E. Griffiths, *J. Food Technol.* **1**, 64 (1964).
35. J. E. McNeal, *J. Assoc. Off. Anal. Chem.* **59**, 570 (1976).
36. S. N. Tewari, S. C. Sharma and K. V. Sharma, *Chromatographia* **7**, 36, 308 (1974).
37. M. H. Peniver, *J. Pharm. Soc.* **57**, 2132 (1968).
38. P. Wollenweber, *J. Chromatogr.* **7**, 557 (1962).
39. G. E. Martin and D. M. Figert, *J. Assoc. Off. Anal. Chem.* **57**, 217 (1974).
40. R. J. Graham and A. E. Nya, *Fifth International Symposium on Chromatography and Electrophoresis*, Ann Arbor Humphrey Science Publications Inc., Michigan, 1969, p. 489.
41. H. Pietsch and R. Meyer, *Nahrung* **9**, 154 (1963).
42. R. Takeshita, T. Yamashita and N. Itoh, *J. Chromatogr.* **73**, 173 (1972).

CHAPTER 5

COSMETICS

5.1 INTRODUCTION

The present-day production of cosmetics has developed into a worldwide multibillion dollar industry with a large range of chemicals being put to a wide variety of different uses, encompassing everything from baby preparations to sun tan and screening formulae.

It is only relatively recently that concern about the possible deleterious side effects of cosmetics has been expressed as a result of the increasing general awareness of the dangers attached to the use of artificial chemicals on humans and the environment. It is probable, therefore, that cosmetic products will soon be monitored, controlled and evaluated for properties such as oral toxicity, inhalation toxicity, phototoxicity, ease of systemic absorption, ophthalmic irritancy, dermal sensitization, carcinogenesis, mutagenesis, and tetratogenesis, as are drugs at present.

The Federal Food, Drug and Cosmetic Act of 1938 was the first attempt by the United States to control cosmetic products. A cosmetic as defined by the Act is 'any article or articles intended to be rubbed, poured, sprinkled or sprayed on, introduced into, or otherwise applied, to the human body or any part thereof for cleansing, beautifying, promoting attractiveness, or altering the appearance, and articles intended for use as components of such an article.' However, even more stringent control has now been requested for cosmetic chemicals.

The European Economic Community has also become concerned with the quality and safety of cosmetic products, as their import and export potential has increased and with it the importance of the identification and analysis of their ingredients.

In order to give an idea of the variety of chemicals used in various cosmetics, some typical examples are cited below and their ingredients listed.

5.1.1 A deodorant antiperspirant

Chemical components: aluminum chlorohydrate, urea, glycerol stearate, magnesium aluminum silicate, PEG 40 stearate, mineral oil, Fragrance, hydroxyethyl cellulose, tetrasodium EDTA, FD and C Blue No. 1, D and C Red No. 19.

5.1.2 A hair shampoo

Chemical components: ammonium lauryl sulfate, lauramide DEA, DEA-linoleate, stearamine oxide, laurol sarcosinen nonoxynol-12, PEG-14M, canada balsam, hydroxypropyl methyl cellulose, trisodium HEDTA, Fragrance, UV absorber 1, methyl paraben, formaldehyde, FD and C Yellow No. 5, FD and C Red No. 4, D and C Red No. 33, FD and C Blue No. 1.

5.1.3 A skin moisturizer

Chemical components: mineral oil, propylene glycol, glycerol stearate, stearic acid, lanolin oil, propyl paraben, lanolin alcohol, oat flour, methyl paraben, triethanolamine, carbomer-341, Fragrance, FD and C Yellow No. 5, D and C Red No. 19.

It can be seen that a very wide variety of compounds find their way into cosmetic formulations. Some chemicals are not permitted in quantities above a certain percentage of the cosmetic, though some users still experience an allergic reaction despite limitations on the levels of allergenic compounds used.

Paper and thin layer chromatography can be used to assay for many of these constituent compounds, although no extraction and isolation procedures have been developed specifically for cosmetic analysis. However, methods and techniques used for food, water and pharmaceutical analysis can be adapted to this purpose.

5.2 ALIPHATIC ACIDS

Aliphatic acids commonly used in cosmetic formulations have been separated by a variety of techniques from matrices other than cosmetic preparations.

Gütlbauer[1] converted the acids to diethylamine salts, dissolved them in methanol and then spotted the solution onto a cellulose thin layer plate.

Adsorbent: Cellulose powder HR 300

Solvent: *n*-butanol–diethylamine–water (85:1:14). The plates were equilibrated with solvent mixture vapors for 4 h

R_F: formic acid = 0.41; acetic acid = 0.45; caprylic acid = 0.95; glycolic acid = 0.25; lactic acid = 0.31; glyceric acid = 0.12; chloracetic acid = 0.52; glyoxylic acid = 0.30; thioglycolic acid = 0.52.

The spots were detected with ninhydrin.

Knappe and Peteri[2] impregnated Silica Gel G plates with Polyethylene Glycol M 1000 (1 part gel + 0.5 part glycol). The plate was then developed with a solvent mixture of di-isopropyl ether, formic acid and water (90:7:3) which was also saturated with the glycol. R_F values obtained were: succinic acid = 0.28; adipic acid = 0.43; malic acid = 0.20. Spots were detected with an acid molybdate solution.

Silica Gel H was used by Ting and Dugger[3] in conjunction with a solvent system of water-saturated diethyl ether and formic acid (7:1). R_F values obtained were: malic acid = 0.53; citric acid = 0.37; tartaric acid = 0.23. Roomi *et al.*[4] used silica gel with both adsorption and reverse phase techniques. The systems used in the adsorption technique were as follows.

Adsorbent: Silica Gel G

Solvent: petroleum ether–diethyl ether (95:5)

R_F: oleic acid = 0.41; behenic acid = 0.52; stearic acid = 0.45; palmitic acid = 0.44.

The reverse phase technique used the following systems.

Adsorbent: Silica Gel G impregnated with DC 200 silicon oil

Solvent: acetonitrile–acetic acid–water (70:10:20)

R_F: oleic acid = 0.50; linoleic acid = 0.62; behenic acid = 0.03; stearic acid = 0.40; palmitic acid = 0.50; lauric acid = 0.71.

Lynes[5] used silica gel with a solvent system of methyl acetate and ammonium hydroxide 2.5% (95:5). The plate was developed twice using fresh solvent for each run, and the resulting R_F values were: formic acid = 0.05; acetic acid = 0.10; propionic acid = 0.15; *n*-butyric acid = 0.24; *n*-valeric acid = 0.39; *n*-hexanoic acid = 0.52; *n*-heptanoic acid = 0.55; *n*-octanoic acid = 0.58; *n*-nonanoic acid = 0.61. The spots were made visible using a methyl red indicator.

5.3 CARBOXYLIC ESTERS

Some carboxylic esters are used as softeners, and Braun[6] chromatographed a number of these using the following system.

Adsorbent: Silica Gel G

Solvent system: methylene chloride

R_F: dioctyl adipate = 0.49; acetyl tributyl citrate = 0.32; acetyl triethyl citrate = 0.19; acetyl tri-2-ethylhexyl citrate = 0.52; dibutyl phthalate = 0.52; diethyl phthalate = 0.38; dinonyl phthalate = 0.60; dioctyl phthalate = 0.59; dioctyl sebacate = 0.35.

5.4 FRAGRANCES

Synthetic perfume-making began in 1837 when Liebig and Wohler manufactured benzaldehyde, a constituent of bitter almond oil. Since then, such essences have been used not only in perfumes and toilet waters but in most cosmetic preparations, as well as in a number of food products. They are, as a rule, chemicals that stimulate the sense of smell in a pleasant manner, and include terpenes, alcohols, esters, aldehydes, coumarins, ionones and musks as well as compounds with other basic chemical structures.

Kirchner and Miller[7] carried out some of the pioneer work in the thin layer chromatography of the terpene essences. They prepared thin layer plates using silicic acid as adsorbent and a 5% starch binder. The plate was dried over phosphorus pentoxide. Hexane was used as the developing solvent. The R_F values obtained were camphene = 0.74; cedrene = 0.82; cumene = 0.62; limonene = 0.41; myrcene = 0.56; α-pinene = 0.83; β-pinene = 0.80; terpinolene = 0.64.

Miller and Kirchner[8] prepared silicic acid–starch binder plates without drying over phosphorus pentoxide, and developed with two different solvents.

Solvent 1: hexane

R_F: camphene = 0.84; α-caryophyllene = 0.50; β-caryophyllene = 0.62; γ-caryophyllene = 0.80; cedrene = 0.82; limonene = 0.41; α-pinene = 0.83; β-pinene = 0.80; terpinolene = 0.64.

Solvent 2: cyclohexane

R_F: camphene = 0.79; α-caryophyllene = 0.47; β-caryophyllene = 0.52; γ-caryophyllene = 0.87; cedrene = 0.83; limonene = 0.54; α-pinene = 0.84; β-pinene = 0.80; terpinolene = 0.64.

Some aldehyde essences were separated by Sundt and Saccardi.[9]

Adsorbent: Silica Gel G

Solvent system A: petroleum ether–ethyl acetate (5:2.5)

R_F: benzylvanillin = 0.61; coumarin = 0.50; dihydrocoumarin = 0.62; 2,4-dimethoxybenzaldehyde = 0.50; 3-ethylcoumarin = 0.81; 4-hydroxy-3-ethoxybenzaldehyde = 0.46; *o*-hydroxybenzaldehyde = 0.46; *p*-hydroxybenzaldehyde = 0.32; *o*-methoxybenzaldehyde = 0.70; *p*-methoxybenzaldehyde = 0.64; 3-methylcoumarin = 0.68; 6-methylcoumarin = 0.55; 3,4-dimethoxybenzaldehyde = 0.41; piperonal = 0.66; propenylguanethol = 0.81; vanillin = 0.30; *o*-vanillin = 0.39.

Solvent system B: hexane–ethyl acetate (5:2)

R_F: benzylvanillin = 0.46; coumarin = 0.38; dihydrocoumarin = 0.50; 2,4-dimethoxybenzaldehyde = 0.44; 3-ethylcoumarin = 0.75; 4-hydroxy-3-ethoxybenzaldehyde = 0.44; 4-hydroxy-3-ethoxybenzaldehyde = 0.42; *m*-hydroxybenzaldehyde = 0.37; *p*-hydroxybenzaldehyde = 0.27; *o*-methoxybenzaldehyde = 0.62; *p*-methoxybenzaldehyde = 0.53; 3-methylcoumarin = 0.58; 6-methylcoumarin = 0.43; 3,4-dimethoxybenzaldehyde = 0.38; piperonal = 0.60; 1-ethoxy-2-hydroxy-4-propenylbenzene = 0.74; 4-hydroxy-3-methoxybenzaldehyde = 0.27; *o*-vanillin = 0.27.

Diazotized sulfanilic acid was used to detect the coumarins as orange spots, whereas hydrazine was required to detect all the other compounds.

Mehlitz *et al.*[10] converted some aromatic essences to their 2,4-dinitrophenylhydrazone derivatives and then spotted them onto a Silica Gel G plate and developed with hexane, carbon tetrachloride and ethyl acetate (2:10:1). The R_F values obtained are given here relative to the R_F of Butter Yellow which was taken as 1.00. Acetone = 0.61; acetophenone = 0.90; acrolein = 0.62; anisaldehyde = 0.50; benzaldehyde = 0.76; *n*-butanal = 0.74; D-carvone = 1.15; L-carvone = 1.13; citral = 0.98, 1.05; crotonaldehyde = 0.67; decanal = 0.75; dodecanal = 1.07; 2-decen-l-al = 1.06; ethanal = 0.46; furfural = 0.43; heptanal = 0.93; hexanal = 0.88; α-ionone = 1.20; β-ionone = 1.11; isobutanal = 0.82; isopentanal = 0.82; nonanal = 0.99; propanal = 0.64. The spots were rendered visible by spraying the chromatogram with a solution of 0.2% potassium ferrocyanide and 0.1% ferric chloride in 2N hydrochloric acid.

Dhant and Mulders-Dijkman[11] also made 2,4-dinitrophenylhydrazone derivatives before spotting, but they used multiple development techniques.

Adsorbent: Silica Gel G (equilibrated in laboratory atmosphere humidity)

Solvent: hexane–benzene (1:1) (five multiple developments)

R_F: butanal = 0.39; butan-2-one = 0.37; carvone = 0.62; carvonane acetone = 0.67; citral = 0.38; citronellal = 0.49; cycloheptanone = 0.39; cyclohexanone = 0.35; cyclooctanone = 0.30; cyclopentanone = 0.25; decanal = 0.57; dihydrocarvone = 0.57; dodecanal = 0.60; ethanol = 0.22; heptanal = 0.51; hexanal = 0.48; heptan-2-one = 0.50; hexan-2-one = 0.46; α-ionone = 0.50; β-ionone = 0.63; isomenthone = 0.66; isopulgenone = 0.52; methanal = 0.23; menthone = 0.51; methylheptenone = 0.44; α-methylionone = 0.67; β-methylionone = 0.66; nonanal = 0.55; nonan-2-one = 0.55; octanal = 0.53; octan-2-one = 0.41; perillaldehyde = 0.44; α-phellandral = 0.64; piperidone = 0.39; propanal = 0.33; propan-2-one = 0.22; *pseudo*-ionone = 0.51; pulgenone = 0.47; α-thujone = 0.53; β-thujone = 0.63; undecanal = 0.59; undecan-2-one = 0.60.

Adsorbent: Silica Gel G (impregnated with adsorbed nitromethane vapors)

Solvent: hexane (three multiple developments)

R_F: butanal = 0.10; butan-2-one = 0.13; carvone = 0.32; carvone acetone = 0.49; citral = 0.23; citronellal = 0.32; cycloheptanone = 0.24; cyclohexanone = 0.17; cyclooctanone = 0.32; cyclopentanone = 0.11; decanal = 0.57; dihydrocarvone = 0.34; dodecanal = 0.76; ethanal = 0.05; heptanal = 0.27; hexanal = 0.21; heptan-2-one = 0.34; hexan-2-one = 0.26; α-ionone = 0.53; β-ionone = 0.60; isomenthone = 0.54; isopulgenone = 0.41; methanal = 0.02; menthone = 0.55; methylheptenone = 0.28; α-methylionone = 0.68; β-methylionone = 0.73; nonal = 0.46; nonan-2-one = 0.55; octanal = 0.37; octan-2-one = 0.44; pentanal = 0.15; pentan-2-

one = 0.19; perillaldehyde = 0.20; α-phellandral = 0.50; piperidone = 0.43; propanal = 0.07; propan-2-one = 0.07; *pseudo*-ionone = 0.40; pulegone = 0.57; α-thujone = 0.46; β-thujone = 0.46; undecanal = 0.66; undecan-2-one = 0.73.

Adsorbent: Silica Gel G (equilibrated in acetonitrile vapors)

Solvent: hexane–benzene (1:1) (five developments in a saturated tank)

R_F: butanal = 0.24; butan-2-one = 0.28; carvone = 0.54; citral = 0.44; citronellal = 0.52; cycloheptanone = 0.47; cyclohexanone = 0.37; cyclooctanone = 0.53; cyclopentanone = 0.28; decanal = 0.73; dihydrocarvone = 0.57; dodecanal = 0.84; ethanal = 0.11; heptanal = 0.47; hexanal = 0.38; heptan-2-one = 0.53; α-ionone = 0.73; β-ionone = 0.76; hexan-2-one = 0.45; isomenthone = 0.73; isopulgenone = 0.56; menthanal = 0.06; menthone = 0.73; methylheptenone = 0.48; α-methylionone = 0.79; β-methylionone = 0.82; nonanal = 0.65; nonan-2-one = 0.70; octanal = 0.56; octan-2-one = 0.63; pentanal = 0.31; pentan-2-one = 0.63; perillaldehyde = 0.39; α-phellandral = 0.68; piperidone = 0.64; propanal = 0.17; propan-2-one = 0.18; *pseudo*-ionone = 0.60; pulgenone = 0.64; α-thujone = 0.66; β-thujone = 0.66; undecanal = 0.78; undecan-2-one = 0.82.

The aldehyde and ketone essences were also chromatographed without derivatization by Klouwen *et al.*[12]

Adsorbent: Silica Gel G

Solvent 1: benzene (saturated development)

R_F: citral = 0.13; citronellal = 0.31; cyclamenaldehyde = 0.37; safranal = 0.10; furfuraldehyde = 0.13; benzaldehyde = 0.32; *p*-tolualdehyde = 0.27; cuminaldehyde = 0.33; α-phenylpropionaldehyde = 0.35; β-phenylpropionaldehyde = 0.30; cinnamaldehyde = 0.19; α-amyl cinnamaldehyde = 0.33; α-hexyl cinnamaldehyde = 0.36; *o*-anisaldehyde = 0.21; *p*-anisaldehyde = 0.18; piperonal = 0.18. Ketone R_F values were: camphor = 0.06; carvone = 0.13; dihydrocarvone = 0.14; zingerone = 0.03; menthone = 0.24; α-thujone = 0.29; β-thujone = 0.25; acorone = 0.03; phyllantone = 0.18; methyl naphthyl ketone = 0.18; acetophenone = 0.17; methoxyacetophenone = 0.10; benzophenone = 0.31.

Solvent 2: benzene–ethyl acetate–acetic acid (90:5:5)

R_F: citral = 0.42; citronellal = 0.65; cyclamenaldehyde = 0.63; benzaldehyde = 0.56; *p*-tolualdehyde = 0.50; cuminaldehyde = 0.54; β-phenylpropionaldehyde = 0.52; cinnamaldehyde = 0.45; α-amylcinnamaldehyde = 0.63; α-hexylcinnamaldehyde = 0.64; salicylaldehyde = 0.48; *m*-hydroxybenzaldehyde = 0.21; *p*-hydrobenzaldehyde = 0.16; syringaldehyde = 0.18; protocatechaldehyde = 0.05; β-resorcylaldehyde = 0.21; gentisaldehyde = 0.19; *o*-vanillin = 0.27; *o*-ethyl vanillin = 0.49; isovanillin = 0.20; ethylisovanillin = 0.25; *p*-vanillin = 0.27; *p*-ethylvanillin = 0.32; *o*-anisaldehyde = 0.51; *m*-methoxybenzaldehyde = 0.55; *p*-anisaldehyde = 0.45; 2,3-dimethoxybenzaldehyde = 0.48; 2,4-dimethoxybenzaldehyde = 0.38; 2,5-dimethoxybenzaldehyde = 0.49; 3,5-dimethoxybenzaldehyde = 0.55; piperonal = 0.45; 3,4,5-trimethoxybenzaldehyde = 0.34; 2,4,5-trimethoxybenzaldehyde = 0.21. The R_F values for the ketones were: camphor = 0.33; carvone = 0.48; dihydrocarvone = 0.48; zingerone = 0.26; menthone = 0.62; α-thujone = 0.62; β-thujone = 0.62; acorone = 0.35; isoacorone = 0.37; frambinone = 0.21; phyllantone = 0.60; methyl naphthyl ketone = 0.59; acetophenone = 0.62; methoxy acetophenone = 0.38.

Solvent 3: chloroform

R_F: citral = 0.51; citronellal = 0.64; hydroxycitronellal = 0.11; cyclamenaldehyde = 0.65; furfuraldehyde = 0.50; benzaldehyde = 0.59; *p*-tolualdehyde = 0.63; cuminaldehyde = 0.65; β-phenylpropionaldehyde = 0.55; cinnamaldehyde = 0.52; α-amylcinnamaldehyde = 0.66; α-hexylcinnamaldehyde = 0.65; salicylaldehyde = 0.57; syringaldehyde = 0.19; *o*-vanillin = 0.47; *o*-ethyl vanillin = 0.56; isovanillin = 0.21; ethylisovanillin = 0.29; *p*-vanillin = 0.27; *p*-ethylvanillin = 0.38; *o*-anisaldehyde = 0.57; *m*-methoxybenzaldehyde = 0.60; *p*-anisaldehyde = 0.56; 2,3-dimethoxybenzaldehyde = 0.57; 2,4-dimethoxybenzaldehyde = 0.50; 2,5-dimethoxybenzaldehyde = 0.58; 3,5-dimethoxybenzaldehyde = 0.64; piperonal = 0.57; 3,4,5-trimethoxybenzaldehyde = 0.41; 2,4,5-trimethoxybenzaldehyde = 0.40. The ketone R_F values were: camphor = 0.27; carvone = 0.63; dihydrocarvone = 0.65; zingerone = 0.30; fenchone = 0.33; piperitone = 0.45; α-thujone = 0.63; β-thujone = 0.64; acorone = 0.47; isoacorone = 0.48; methyl naphthyl ketone = 0.59; acetophenone = 0.62; methoxyacetophenone = 0.55.

The compounds were detected by viewing the thin layer plate under UV illumination before and after treatment with a hydrazinium sulfate spray.

Klouwen and Ter-Heide[13] were also able to separate the phenolic fragrances by thin layer chromatography with two different solvent systems.

Adsorbent: Silica Gel G

Solvent A: chloroform

R_F (referenced to thymol = 1.00): anethol = 1.66; carvacrol = 0.94; chavicol methyl ether = 1.66; cresol = 1.07; dihydro-eugenol = 1.20; eugenol = 1.24; eugenol benzyl ether = 1.53; eugenol methyl ether = 1.48; guaiacol = 1.13, 1.24; *p*-hydroxyanisole = 0.37; isochavibetol = 1.18; iso-eugenol = 1.20; iso-eugenol benzyl ether = 1.65; iso-eugenol methyl ether = 1.52; isomyristicin = 1.58; isosafrol = 1.66; isothymol = 0.67; 3,4-methylenedioxyphenol = 0.42; myristicin = 1.54; α-naphthol = 0.67; β-naphthol = 0.52; orcin = 0.03; phenol = 0.49; safrol = 1.64.

Solvent B: petroleum ether–acetic acid (95:5)

R_F (referenced to thymol = 1.00): anethol = 4.93; carvacrol = 1.10; chavicol methyl ether = 5.00; cresol = 1.02; dihydro-eugenol = 1.98; eugenol = 1.37; isochavibetol = 1.33; iso-eugenol = 1.23; iso-myristicin = 1.91; isosafrol = 5.06; isothymol = 0.63; myristicin = 1.87; α-naphthol = 0.43; β-naphthol = 0.32; phenol = 0.47; safrol = 5.45.

The compounds were detected in all cases by a carbon tetrachloride solution of antimony pentachloride.

5.5 PRESERVATIVES

Preservatives are widely used in cosmetic products and are usually present in concentrations of less than 0.1%. Wilson[14] used precoated Silica Gel GF_{254} for thin layer chromatography in conjunction with two solvent systems.

Solvent A: benzene–acetone (8:2)

R_F: 2-bromo-2-nitropropane-1,3-diol = 0.47; butylated hydroxy-toluene = 0.79; isopropyl-*o*-cresol = 0.65; chlorhexidine acetate = 0.40; 4-chloro-3-methylphenol = 0.59; *p*-chloro-*m*-xylenol = 0.60; dehydroacetic acid = 0.15; dichloro-*m*-xylenol = 0.65; dichlor-pene = 0.50; 3,5-dibromo-3′-trifluoromethylsalicylamide = 0.56; hexachlorophene = 0.14; trichlosan = 0.74; methyloyldimethyl-hydantoin = 0.23; octyl gallate = 0.12; *o*-phenylphenol = 0.68; pro-pylparaben = 0.56; resorcinol = 0.39; salicylanilide = 0.65; tetra-bromo-*o*-cresol = 0.72; 3,4′,5-tribromosalicylanilide = 0.60; 3,4,4′-trichlorocarbanilide = 0.55; *N*-trichloromethylthio-4-cyclohexane-1,2-dicarboximide (captan) = 0.70; zinc pyrithone = 0.49.

Solvent B: chloroform–methanol (9:1)

R_F: Bronopol = 0.57; butylated hydroxytoluene = 0.83; carvacrol = 0.75; chlorhexidine acetate = 0.62; 4-chloro-3-methylphenol = 0.65; *p*-chloro-*m*-xylenol = 0.67; dehydro-acetic acid = 0.26; dichloro-*m*-xylenol = 0.75; fluorsalan = 0.53; hexachlorophene = 0.41; Irgasan CF = 0.67; Irgasan DP = 0.81; MDM hydantoin = 0.59; octyl gallate = 0.22; *o*-phenylphenol = 0.79; propylparaben = 0.65; resorcinol = 0.34; salicylanilide = 0.72; tetrabromo-*o*-cresol = 0.80; tribromsalan = 0.69; trichlocarban = 0.70; captan = 0.83; zinc pyrithone = 0.82.

The compounds were detected with an aminoantipyrine and potassium ferri-cyanide mixture, or with diazotized benzidine, or by treatment with 4-methyl umbelliferone after exposure to iodine vapors.

5.6 MISCELLANEOUS COSMETIC COMPONENTS

5.6.1 Paper chromatographic separation

Barker and Sawicki[15] chromatographed some aldehydes with immobile–mobile phase chromatography.

Paper: Schleicher and Schuell 2043b
Immobile phase: 35% formamide in ethanol
Mobile phase: cyclohexane–benzene–dipropylene glycol (70:30:3)

R_F: acetophenone = 0.56; anisaldehyde = 0.26; benzaldehyde = 0.44; benzophenone = 0.94; cinnamaldehyde = 0.60; *o*-methoxybenzaldehyde = 0.29; piperonal = 0.17.

Immobile phase: 50% *N,N*-dimethylformamide
Mobile phase: dibutyl ether–*N,N*-dimethylformamide–tetrahydrofuran (85:15:4)

R_F: acetophenone = 0.45; anisaldehyde = 0.26; benzaldehyde = 0.44; benzophenone = 0.58; cinammaldehyde = 0.43; furfural = 0.33; *m*-hydroxybenzaldehyde = 0.24; *p*-hydroxybenzaldehyde = 0.23; 2-hydroxy-3-methoxybenzaldehyde = 0.20; *o*-methoxybenzaldehyde = 0.30; piperonal = 0.24; salicylaldehyde = 0.34.

Alcohols and glycols were chromatographed by Brownell.[16]

Chromatographic paper: Whatman No. 1
Solvent A: heptane saturated with methanol

R_F: cinnamyl alcohol = 0.25; octanol = 0.64; benzoin = 0.23; benzhydrol = 0.34; cholesterol = 0.90.

Solvent B: *n*-butanol saturated with water.

R_F: diethylene glycol − 0.62; ethylene glycol = 0.52; triethylene glycol = 0.63; glycerol = 0.35.

To render the spots visible the chromatogram was steamed and then sprayed with a solution of 3,5-dinitrobenzoyl chloride containing rhodamine.

Bergner and Sperlich[17] separated glycols by paper chromatography, with a developing solvent of chloroform and ethanol (8:2), and the R_F values were: ethylene glycol = 0.47; 1,3-propylene glycol = 0.55; 1,2-propylene glycol = 0.65; diethylene glycol = 0.76; triethylene glycol = 0.81.

5.6.2 Thin layer chromatography

Comer and Comer[18] were able to separate some of the basic constituents of cosmetics by thin layer chromatography.

Adsorbent: silica gel
Developing solvent 1: *n*-butanol–5N ammonium hydroxide–methanol (60:20:20)

R_F: polyethylene glycol = 0.49; cocoa butter = 0.80; ethyl *p*-hydroxybenzoate = 0.66; glucose = 0.14; glycerin = 0.33; heavy mineral oil = 0.78; polyoxol 40 stearate = 0.54; sorbitan sesqui-oleate = 0.77; spermaceti = 0.77; magnesium stearate = 0.50.

Developing solvent 2: cyclohexane–acetone (40:50)

R_F: cetyl alcohol = 0.70; ethyl cellulose = 0.80; ethyl *p*-hydroxybenzoate = 0.61; cocoa butter = 0.77.

Developing solvent 3: chloroform–ethanol (9:1)

R_F: cetyl alcohol = 0.70; cocoa butter = 0.81; ethyl cellulose = 0.71;

ethyl *p*-hydroxybenzoate = 0.65; light paraffin oil = 0.77; polyethylene glycol 200 = 0.07; spermaceti = 0.81; white petrolatum = 0.80; white wax = 0.82.

The migrated spots were detected with iodine vapors, or with solution of potassium permanganate oversprayed with sulfuric acid in the case of glycerin and mineral oil.

5.7 HAIR DYES

These compounds can be classified as four general types.

(a) *Color restorers* are metallic dyes capable of building up hair color by repeated applications. They usually contain lead acetate.
(b) *Color rinses* are sold as aqueous solutions of dyes containing a small amount of resin and can give a light coloration to the hair. Temporary hair colorants contain high molecular weight dyes such as textile dyes, FD and C, D and C and Ext. D and C colors.
(c) *Semipermanent hair dyes* are sold in shampoo form and contain dyes of relatively low molecular weight in a simple emulsion system.
(d) *Long lasting hair dyes* are oxidation dyes. They are a mixture of diaminobenzenes, aminophenols and polyhydroxybenzenes in an ammonium oleate soap or an ammonia-containing synthetic surfactant, which must be mixed with hydrogen peroxide or solid oxidizing agents such as urea peroxide and sodium perborate just before application. The color develops inside the hair fiber by the oxidative coupling of *p*-diamines and *p*-aminophenols to quinoneimines which in turn couple with other phenols and amines to form a wide variety of indo-aniline and indamine dyes. The mechanism is quite similar to that which takes place in the coupling dye formation of photographic film.

Food additive dyes have undergone a great deal of scrutiny in the last few years and such is now the case for chemicals used in hair dyeing, for which more and more toxicity studies are now being carried out.

5.7.1 Paper chromatography

Reio[19] used paper chromatography to separate some of the diamines.

Adsorbent: Whatman No. 1 paper

Solvent 1: methyl ethyl ketone–acetone–formic acid–water (40:2:1:6)

R_F: 4-nitro-*o*-phenylenediamine = 0.94; *m*-phenylenediamine = 0.15; *o*-phenylenediamine = 0.74; *p*-phenylenediamine = 0.28.

Solvent 2: methyl ethyl ketone–diethylamine–water (92:0.2:0.8)

R_F: 4-nitro-*o*-phenylenediamine = 0.92; *m*-phenylenediamine = 0.77; *o*-phenylenediamine = 0.83; *p*-phenylenediamine = 0.59.

Solvent 3: benzene–formic acid–water (10 parts of benzene saturated with 1 part of 2% formic acid).

An R_F of 0.23 was obtained for 4-nitro-*o*-phenylenediamine using this solvent system. The compounds were detected under UV radiation, and by coupling either with diazotizd sulfanilic acid or with diazotized *o*-dianisidine.

Smyth[20] separated the oxidative type of hair dyes on Whatman No. 3MM paper with the upper layer of the following mixture used as the developing solvent: *n*-butanol–ethanol–water–acetic acid–sodium sulfite (50:10:40:1:1). R_F values obtained for this system were: 2-amino-1-phenolsulfonic acid = 0.08; 2,4-diaminophenol = 0.34; *p*-phenylenediamine = 0.36; *m*-phenylenediamine = 0.46; 4-methoxy-*m*-phenylenediamine = 0.49; 2,5-diaminotoluene = 0.49; 2,4-diaminotoluene = 0.58; *p*-aminophenol = 0.59; *p*-methylaminophenol = 0.62; *o*-phenylenediamine = 0.63; 2-nitro-*p*-phenylenediamine = 0.64; *p*-diethylaminoaniline = 0.68; *m*-aminophenol = 0.71; *p*-anisidine = 0.72; 4-nitro-*p*-phenylenediamine = 0.73; picramic acid = 0.75; pyrogallol = 0.75; *o*-aminophenol = 0.77; *N*-phenyl-*p*-phenylenediamine = 0.79; 4-amino-2-nitrophenol = 0.82; 4-chloro-*o*-phenylenediamine = 0.83; hydroquinone = 0.86; catechol = 0.82; 2-amino-nitrophenol = 0.89; *o*-anisidine = 0.89; resorcinol = 0.90. The resolved compounds were detected with ammoniacal silver nitrate or by spraying the paper with an ethanolic solution of *p*-dimethylaminobenzaldehyde in hydrochloric acid.

5.7.2 Thin layer chromatography

Corbett and Caterbone[21] resolved a whole series of substituted 2-nitro-*p*-phenylenediamine compounds using Quantagram LQD precoated silica gel plates and multiple development in two separate solvent systems.

1st Developer: methanol–methylene chloride (5:95) (plate developed for 5 cm and then dried)

2nd Developer: methanol–isopropanol–methylene chloride (3:2:95)

R_F: *N*,*N*,*N*′-trimethyl-2-nitro-*p*-phenylenediamine = 0.81; *N*,*N*′-dimethyl-2-nitro-*p*-phenylenediamine = 0.80; *N*,*N*-dimethyl-2-nitro-*p*-phenylenediamine = 0.78; *N*-methyl-2-nitro-*p*-phenylenediamine = 0.76; *N*′-methyl-2-nitro-*p*-phenylenediamine = 0.75; 2-nitro-*p*-phenylenediamine = 0.70; *N*-methyl-*N*-2-hydroxymethyl-*N*′-methyl-2-nitro-*p*-phenylenediamine = 0.68; *N*,*N*-dimethyl-*N*′-2-hydroxymethyl-2-nitro-*p*-phenylenediamine = 0.63; *N*-methyl-*N*-2-hydroxymethyl-2-nitro-*p*-phenylenediamine = 0.60; *N*-methyl-*N*′-2-hydroxymethyl-2-nitro-*p*-phenylenediamine = 0.59; *N*-2-hydroxymethyl-*N*′-methyl-2-nitro-*p*-phenylenediamine = 0.58; *N*′-2-hydroxymethyl-2-nitro-*p*-phenylenediamine = 0.48; *N*-2-hydroxymethyl-2-nitro-*p*-phenylenediamine = 0.47; *N*,*N*-2-hydroxymethyl-*N*′-methyl-2-nitro-*p*-phenylenediamine = 0.42; *N*-methyl-*N*-2-hydroxymethyl-*N*′-2-

hydroxymethyl-2-nitro-*p*-phenylenediamine = 0.38; *N*,*N*-2-hydroxymethyl-2-nitro-*p*-phenylenediamine = 0.33; *N*-2-hydroxymethyl-*N*′-2-hydroxymethyl-2-nitro-*p*-phenylenediamine = 0.28; *N*,*N*-2-hydroxymethyl-*N*′-2-hydroxymethyl-2-nitro-*p*-phenylenediamine = 0.16.

After being resolved, the compounds were made visible by coupling with *p*-dimethylaminobenzaldehyde under acidic conditions to give colored anils, or with diazotized *p*-nitroaniline which will couple to give diazo dyes, or with ammoniacal silver nitrate which will be reduced to form gray spots where it interacts with a dye.

Although government departments do not at the present time test cosmetic dyes as they do food dyes, the adsorbent and solvent systems used in their routine identification tests with food colorings can also be used for cosmetic colorings. Peniver[22] was able to chromatograph drug and cosmetic dyes on cellulose layers.

Adsorbent: cellulose

Solvent system 1: ethyl acetate–*n*-butanol–water–pyridine (5:5:5:6)

R_F: D and C No. 5 (Alizarin Cyanine Green F) = 0.64; D and C No. 6 (Quinizarin Greeen SS) = 1; D and C No. 4 (Orange II) = 0.76; D and C No. 5 (Dibromofluorescein) = 0.76, 0.94; D and C No. 11 (Lithol Red CA) = 0.93; D and C No. 22 (Eosin YS) = 0.88; D and C No. 28 (Phloxine B) = 0.79; D and C No. 33 (Naphthalene Red B) = 0.43; D and C No. 10 (Quinoline Yellow WS) = 0.62, 0.70; D and C No. 11 (Quinoline Yellow SS) = 0.95.

Solvent system 2: ethyl acetate–*n*-butanol–ammonium hydroxide (d = 0.880) (20:55:25)

R_F: D and C No. 5 = 0.38; D and C No. 6 = 1; D and C No. 4 = 0.55; D and C No. 5 = 0.28, 0.23; D and C No. 11 = 0.96; D and C No. 22 = 0.37; D and C No. 28 = 0.51; D and C No. 33 = 0.12; D and C No. 10 = 0.29.

Solvent system 3: *n*-propanol–ethyl acetate–ammonium hydroxide (d = 0.880) (65:75:60)

R_F: D and C No. 5 = 0.63; D and C No. 4 = 0.79; D and C No. 5 = 0.55; D and C No. 22 = 0.61; D and C No. 28 = 0.74; D and C No. 33 = 0.39; D and C No. 10 = 0.58, 0.69.

Other systems described in the chapter on food additives can also be applied to cosmetic dyes.

5.8 PHOTOTOXICITY. IDENTIFICATION OF BERGAPTEN

The possibility of allergic reactions to cosmetics has already been mentioned. Perfume-induced phototoxicity is an example of such allergies. Bergamot oil is frequently used in perfumes, toilet waters and colognes and as a fragrance in some cosmetic preparations. Before 1970 frequent reports were made of skin dermatitis resulting from the use of certain perfumes, which was exacerbated by exposure to sunlight.

Yates *et al.*[23] were able to separate coumarins and psoralens from bergamot oil by preparative thin layer chromatography and by subsequently extracting particular zones for phototoxicity testing. The coumarin–psoralen fraction was initially separated from a sample of bergamot oil by collecting the volatile fraction that distilled at 35–65 °C at a pressure of 2 mm. A sample of 70 mg was spotted on a thin layer Silica Gel G plate (500 μ in thickness). The plates were then developed with toluene, ethyl acetate and acetic acid (80:20:2).

The four zones produced by developing the chromatogram were scraped off for toxicity studies. Zone 1 was contained in the R_F perimeter of 0.64–0.74 and consisted of bergamotin and a small amount of methoxygeranoxycoumarin. Zone 2 occurred in the R_F perimeter of 0.51–0.64 and consisted of methoxygeranoxycoumarin, plus an unidentified material which gave a faint blue fluorescence under UV radiation. Zone 3 was contained in the area of R_F 0.40–0.51 and contained limettin and bergapten (5-methoxypsoralen). Zone 4 contained bergaptol and some unknown compounds and occurred in the area between the origin and R_F 0.40. Testing the fractions, after purification of the extracts, showed that pure bergapten from zone 3 was phototoxic at a level of 20 ppm (20 μg/g).

A method for extracting and isolating bergapten from cosmetic formulations was developed by Wisneski.[24] A 10 ml sample was introduced into a 100 ml beaker and 3 ml of ammonium hydroxide was added with stirring. A 15 ml volume of 1N sodium hydroxide was then added, followed by 95% ethanol until the solution appeared homogeneous. The solution was transferred to a separating funnel, and the beaker was rinsed with 100 ml of water which was also added to the funnel. The solution was then extracted with three separate 50 ml volumes of ethyl ether. Solid sodium sulfate was added to break up any emulsions that might form.

The combined ether extracts were further washed with three separate 25 ml volumes of water. All the aqueous extracts were combined and acidified with hydrochloric acid until strongly acid to litmus. This solution was then extracted with three separate 50 ml volumes of ethyl ether; the ether extracts were again combined and washed with two 50 ml volumes of water and then with three 10 ml volumes of dilute ammonium hydroxide (1:50). Finally, the ether solution was dried over sodium sulfate and was then ready for chromatographic identification.

Swain[25] used Whatman No. 1 chromatographic paper with several different solvent systems for identifying bergapten.

Solvent 1: *n*-butanol–acetic acid–water (4:1:5) (aqueous layer used)
Solvent 2: 1-benzyl alcohol–acetic acid–water (4:1:5)
Solvent 3: 10% aqueous acetic acid
Solvent 4: isopropanol–water (1:4)

R_F = 0.65, 0.59, 0.45 and 0.38, respectively.

5.9 IDENTIFICATION OF NATURAL COUMARINS

Some chromatographic methods can be used to identify natural coumarins that may occur in cosmetic preparations. Stahl and Schorn[26] used Silica Gel G with a developing solvent of toluene, ethyl formate and formic acid (50:40:10). R_F values obtained were: aesculitin = 0.46; nordalbergin = 0.58; umbelliferone = 0.59; marmesin = 0.56; dalbergin = 0.64; xanthotoxol = 0.63; psoralene = 0.69; bergapten = 0.68; xanthotoxin = 0.65; isopimpinellin = 0.67; *o*-methyldalbergin = 0.69; coumarin = 0.71; herniarin = 0.70; imperatorin = 0.73; osthol = 0.78. Abdel-Hay *et al.*[27] also used Silica Gel G, but a solvent mixture of benzene and acetone (9:1) was used for resolution. R_F values in this case were: umbelliferone = 0.27; marmesin = 0.18; dalbergin = 0.40; xanthotoxol = 0.22; psoralene = 0.75; bergapten = 0.75; xanthotoxin = 0.71; isopimpinellin = 0.70; *o*-methyldalbergin = 0.68; coumarin = 0.78; herniarin = 0.74; imperatorin = 0.79; osthol = 0.82.

Karlsen *et al.*[28] used aluminum oxide as adsorbent instead.

Adsorbent: Aluminum Oxide G (activated at 110 °C for 6 h)

Solvent for 1st dimension chromatography: chloroform

R_F: psoralene = 0.51; bergapten = 0.43; xanthotoxin = 0.33; isopimpinellin = 0.20; angelicin = 0.60; isobergapten = 0.58; sphondin = 0.47; pimpinellin = 0.54.

Solvent for 2nd dimension development: dibutyl ether–ethyl acetate (88:12)

R_F: psoralene = 0.46; bergapten = 0.51; xanthotoxin = 0.40; isopimpinellin = 0.42; angelicin = 0.56; isobergapten = 0.60; sphondin = 0.33; pimpinellin = 0.54.

Compounds resolved by all the above systems were detected either by their fluorescence under black light illumination, or by spraying with a solution of 0.2 g iodine and 0.4 g potassium iodide in 100 ml water.

REFERENCES

1. F. Gütlbauer, *J. Chromatogr.* **45**, 104 (1969).
2. E. Knappe and D. Peteri, *Z. Anal. Chem.* **188**, 184 (1962).
3. I. P. Ting and W. M. Dugger Jr, *Anal. Biochem.* **12**, 574 (1965).
4. M. W. Roomi, M. R. Subbaram and K. T. Achaya, *J. Chromatogr.* **15**, 106 (1964); *idem, ibid.* **16**, 295 (1965).
5. A. Lynes, *J. Chromatogr.* **15**, 108 (1964).
6. D. Braun, *Kunstoffe-Plastics* **52**, 2 (1962).
7. J. G. Kirchner and J. M. Miller, *Ind. Eng. Chem.* **44**, 318 (1952).
8. J. M. Miller and J. G. Kirchner, *Anal. Chem.* **25**, 1107 (1953).
9. E. Sundt and A. Saccardi, *Food Technol.* **16**, 89 (1962).
10. A. Mehlitz, K. Gierschner and T. Minas, *Chem. Zeit.* **87**, 573 (1963).
11. J. H. Dhant and G. J. C. Mulders-Dijkman, *Analyst* **94**, 1093 (1969).
12. M. H. Klouwen, R. Ter-Heide and J. G. Kok, *J. Fette Seifen Anstrichm.* **65**, 414 (1963).
13. M. H. Klouwen and R. Ter-Heide, *Parfeum. Kosmet.* **43**, 195 (1962).

14. C. H. Wilson, *J. Soc. Cosmet. Chem.* **26**, 75 (1975).
15. E. D. Barker and E. Sawicki, *Anal. Chem.* **40**, 985 (1968).
16. H. H. Brownell, *J. Chromatogr.* **19**, 556 (1965).
17. K. G. Bergner and H. Sperlich, *Z. Lebensm. Unters. Forsch.* **97**, 253 (1953).
18. J. P. Comer and I. Comer, *J. Pharm. Sci.* **56**, 413 (1967).
19. L. Reio, *J. Chromatogr.* **47**, 60 (1970).
20. R. B. Smyth, *J. Chromatogr.* **16**, 454 (1964).
21. J. F. Corbett and U. Caterbone, *The Analytical Chemistry of Synthetic Dyes*, p. 510, John Wiley, New York (1977).
22. M. H. Peniver, *J. Pharm. Soc.* **57**, 2132 (1968).
23. R. Yates, F. Marzulli and H. I. Maibach, *J. Soc. Cosmet. Chem.* **21**, 695 (1970).
24. H. H. Wisneski, *J. Assoc. Off. Anal. Chem.* **59**, 547 (1976).
25. T. Swain, *Biochem J.* **53**, 200 (1953).
26. E. Stahl and P. J. Schorn, *Z. Physiol. Chem.* **325**, 263 (1961).
27. F. M. Abdel-Hay, E. A. Abu-Mustafa, B. A. El-Tawil and M. B. E. Fayez, *Planta Med.* **13**, 91 (1965).
28. J. Karlsen, L. E. J. Boomsma and A. Svendsen, *J. Chromatogr.* **42**, 550 (1969).

CHAPTER 6

NATURALLY OCCURRING TOXICANTS

Primitive man was probably frequently poisoned in his search for suitable food, since the only way he could determine its suitability was by trial and error. In all likelihood, it was not long before he determined which foods were harmless and how to render foods edible by cooking, peeling, or any other means that he could devise. If this had not been the case, the human species would not have survived.

Although much publicity and news coverage is given to man madc pollutants in the environment which put the human population at risk, little attention is given by the media to the many naturally occurring environmental toxicants which could also be dangerous to human beings if they failed to recognize them as poisons. Most of these natural toxins originate in the food supply that we use for ourselves and our livestock. Indeed, not all the chemicals found in food are beneficial, and though many of these components are metabolized by the liver into nontoxic compounds, those that cannot be thus converted can be damaging to human health.

6.1 GRAYANOTOXINS IN HONEY

Natural food advocates claim that wild honey has greater nutritional value than domestically produced honey. The small nutritional advantage if any, is, however, overshadowed by the possibility that wild honey can contain poisons. Certain plants of the *Ericacaea* family contain grayanotoxins which have a neurotoxic effect. These toxins have the structures shown in Fig. 6.1, and they can occur in the nectar of the flowers, and therefore be stored in the honey produced from them. They can, when ingested by humans, cause nausea, blurred vision, shortness of breath and slight paralysis of the hands and arms.

Chromatography can be used to ascertain the presence or absence of these toxins. Scott *et al.*[1] developed a thin layer chromatographic procedure for isolating and identifying some of these compounds. A homogeneous solution of honey was made by dissolving 50 g in 100 ml of water and methanol (3:1). The pH of the solution was adjusted to 6.5 with dilute sodium hydroxide, then centrifuged and filtered through Whatman No. 40 or other suitable paper. The aqueous methanol solution was extracted with five 150 ml volumes of chloroform, and the combined extracts were evaporated down to a small volume with nitrogen in a steam bath. The residue was transferred to a small container, dissolved in about 0.5 ml of chloroform and spotted onto a thin layer plate for analysis.

Adsorbent: silica gel (Adsorbosil 5) (activated at 105 °C for 2 h)
Solvent system A: benzene–methanol–acetic acid (18:6:1)
Solvent system B: acetone–chloroform (1:1)

With solvent system A the following R_F values were obtained: Grayanotoxin I = 0.37; Toxin II = 0.44; Toxin III = 0.26. Solvent B was used to separate Toxin I from an interference which sometimes appeared.

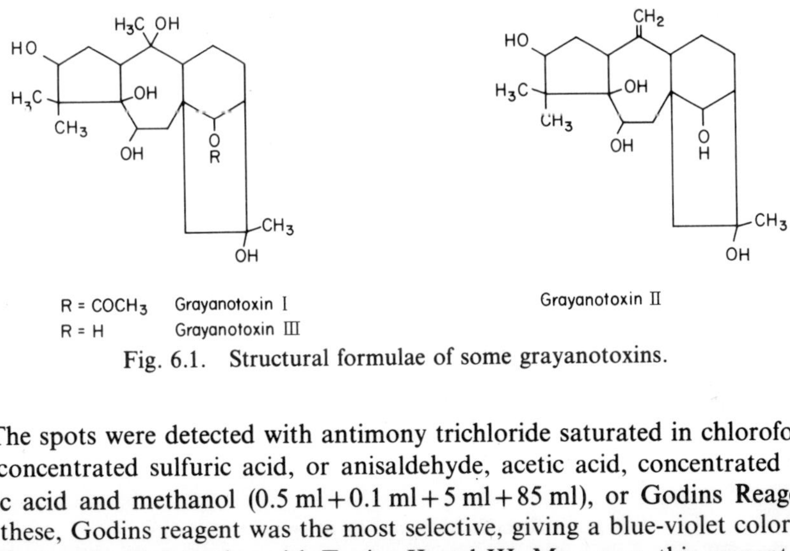

Fig. 6.1. Structural formulae of some grayanotoxins.

The spots were detected with antimony trichloride saturated in chloroform, or concentrated sulfuric acid, or anisaldehyde, acetic acid, concentrated sulfuric acid and methanol (0.5 ml + 0.1 ml + 5 ml + 85 ml), or Godins Reagent. Of these, Godins reagent was the most selective, giving a blue-violet color for Toxin I and a violet color with Toxins II and III. Moreover, this reagent did not give a false color reaction with the interfering material. Further evidence of the presence of the toxins was obtained by scraping the spots off the silica gel plate, extracting the colored material from the silica gel, and injecting the extract into mice to confirm the toxic properties.

6.2 PHALLOTOXINS AND AMATOXINS FROM MUSHROOMS

The eating of wild mushrooms sometimes results in fatalities. The green and white species of the genus *Amanita* which contain two groups of highly toxic polypeptides known as the phallotoxins and the amatoxins, respectively, account for the majority of such poisonings. The former are cyclic heptapeptides while the latter are cyclic octapeptides.

The phallotoxins are fast-acting poisons, while the amatoxins act more slowly (their lethal effects take about 15 hours to develop) though they are about 10 to 20 times more toxic than the phallotoxins. Though it takes a much higher dose of phallotoxins to produce death, a lethal dose will kill in one to two hours. Poisonous species of *Amanita* contain approximately 14 mg of amatoxin per 100 g of the mushrooms. The lethal dose for a human adult is less than 100 μg per kg body-weight, so that the ingestion of a 50 g mushroom could result in death for the adult, whereas a much smaller amount would prove lethal for a child. Figure 6.2 shows the structure of the three polypeptides.

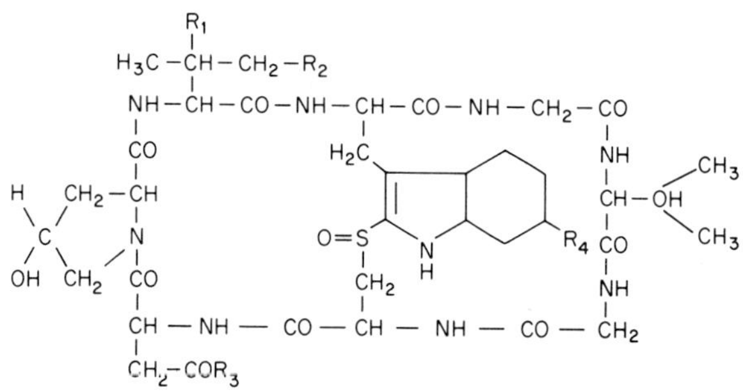

α-Amanitin

$R_1, R_2, R_4 = OH, R_3 = NH_2$

β-Amanitin

$R_3 = OH$

γ-Amanitin

$R_1 = OH$

Fig. 6.2. Structural formulae of some amatoxins.

Palyza[2] developed a thin layer chromatographic method for identifying the amanitins. One gram of mushroom was ground with 5 ml of methanol in a mortar and pestle. The mushroom pulp and solvent were filtered and the filtrate concentrated to a 1 ml volume. This concentrate was then spotted on a thin layer plate.

Adsorbent: Silufol pre-coated silica gel plates (either with 254 A UV indicator or without a UV indicator)

Solvent system: butyl cellosolve–25 % ammonium hydroxide (7:3).

The detection reagent was a 0.2% cinnamaldehyde solution incorporated into the solvent

R_F: α-amanitin = 0.36; β-amanitin = 0.32; γ-amanitin = 0.51.

The developed plate was dried and exposed to hydrochloric acid fumes; the amanitins were detected as violet spots.

Sullivan *et al.*[3] extracted toxins from 4 g of mushrooms with 100 ml of methanol in a Soxhlet apparatus. The extract was chilled to 10 °C and centrifuged, after which the supernatant liquid was decanted, concentrated to 20 ml and chilled again. The chilled extract was centrifuged and the supernatant liquid decanted again. Small amounts of water were added until no further precipitate was visible. The extract was then evaporated to dryness and redissolved in 4 ml of methanol. Aliquots of this solution were spotted on thin layer plates.

Adsorbent: Silica Gel G

Developing solvent: methanol–methyl ethyl ketone (1:1)

R_F: α-amanitin = 0.46; β-amanatin = 0.23.

The compounds were detected by spraying the plates with a 1% methanolic solution of cinnamaldehyde, drying them, and exposing to hydrochloric acid fumes for 10 min. Violet spots were formed.

6.3 TOXINS FROM FISH

Toxins contained in certain species of fish can produce gastric distress and, occasionally, death. Shellfish have been known to contain poisons which cause paralysis and death. The toxins originate in marine dinoflagellates which occur in ocean areas where the shellfish feed; they have been isolated from the hepatopancreas siphon of the shellfish. A poison isolated from the dinoflagellate *Gonyaulax catenella* appears to be identical to the toxin obtained from shellfish.

This form of poisoning occurs around the North Sea Coast, the North Atlantic Coast of America, the North Pacific Coast of America from Central California to the Aleutian Islands and the coastal areas of Japan and South Africa.

Gymnodinium breve is another species of dinoflagellate associated with shellfish poisoning which has been responsible for the mass killing of marine animals.[4] The dinoflagellate was grown in artificial seawater, harvested, the pH adjusted to 5.5 with 1N hydrochloric acid, and the culture was then extracted with diethyl ether. (Each liter of culture was extracted with 150 ml of ether.) The ether extract was evaporated to dryness, and the tarry residue dissolved in acetone and centrifuged. The supernatant liquid was removed and evaporated to dryness. The residue was dissolved in ether and spotted on thin layer plates.

Adsorbent: SilicAR 4GF

Solvent: benzene–ethyl acetate–ethanol–acetic acid (79:10:10:1)

This solvent was used for the first development which resolved twelve bands when viewed under UV radiation. Each band was scraped off, eluted and injected into mice to determine toxic activity. Only the bands with R_F values from 0.41 to 0.49 produced toxic effects. The ether extract of these bands were respotted and developed on a SilicAr 7GF plate in a solvent system of benzene, ethyl acetate and ethanol (80:10:10). This time five bands were visible under UV illumination, but only one with an R_F value of 0.23 to 0.27 proved to be toxic.

This band was scraped off, eluted, and rechromatographed on the SilicAr 7GF with benzene, ethyl acetate and *sec*-butanol (80:10:10). Two bands resulted, one at $R_F = 0.82$ and another at $R_F = 0.69$. Only the material at 0.69 was found to be toxic when it was eluted from the adsorbent. The purified toxin was a yellow-green oil.

It can be seen that thin layer chromatography is useful not only as an identification procedure but also as a purification procedure which enables detection and evaluation.

Oshima *et al.*[5] extracted toxins from clams and species of the algae *Gonyaulax*. Two hundred grams of shucked clams were extracted with 400 ml of 80% ethanol adjusted to pH 2 with hydrochloric acid. The extract was concentrated, washed with chloroform, and chromatographed on a Sephadex G-15 column. Two eluates were obtained, with water and acetic acid as the respective eluting solvents. Final separation was achieved with a column of Biorex 70, eluted with an acetic acid gradient from 0.03N downwards and then with 0.02N hydrochloric acid. Six eluates were collected for chromatographic analysis.

Samples of *Gonyaulax* were collected in 300 ml volumes with the pH adjusted to 2.0, and the samples were then frozen, thawed and centrifuged. The pH of the supernatant liquid was adjusted to 4.2 with 1N sodium hydroxide. It was then evaporated to dryness and extracted twice with 5 ml of 80% ethanol. In preparation for chromatography on isolation columns, the pH of the liquid was adjusted to 6.5 with 0.1N sodium hydroxide.

The six eluates collected for chromatography were given JGX (degree of toxicity to mice) designations.

Adsorbent: Silica Gel 60

Solvent: pyridine–ethyl acetate–water–acetic acid (75:25:30:15)

R_F: JGX1 toxin = 0.61; JGX2 toxin = 0.81; JGX3 toxin = 0.70; JGX4 toxin = 0.61; JGX5 toxin = 0.65; JGX6 toxin = 0.51.

Adsorbent: Silica Gel GF

Solvent: *t*-butanol–acetic acid–water (2:1:1)

R_F: JGX1 toxin = 0.52; JGX2 toxin = 0.65; JGX3 toxin = 0.70; JGX4 toxin = 0.61; JGX5 toxin = 0.65; JGX6 toxin = 0.51.

All the toxins were detected by spraying with 1% hydrogen peroxide and heating at 120 °C for 15 min.

Burke *et al.*[6] isolated toxins from shell fish and from *Gonyaulax catenella.* The toxin was extracted from their tissues by rupturing cell membranes with acid hydrolysis at high temperature. The toxins were separated on an ion-exchange column of XE-64 which had been treated with a pH 5.2 buffer, and were then eluted with the same buffer at 30 drops per min until a 100 ml volume was eluted. The toxin was then eluted with 0.5N trichloracetic acid and collected in a fraction collector. The fractions containing the toxin were determined by bioassay.

Paper chromatography was used for further separation and identification.

Solvent 1: *t*-butyl alcohol–acetic acid–water (2:1:1)
R_F: shellfish toxin = 0.28; *G. catenella* toxin = 0.32.
Solvent 2: *t*-amyl alcohol–pyridine–water (7:7:6)
R_F: shellfish toxin = 0.49; *G. catenella* toxin = 0.51.
Solvent 3: phenol–water (4:1)
R_F: shellfish toxin = 0.29; *G. catenella* toxin = 0.30.
Solvent 4: tetrahydrofurfuryl alcohol–water (17:3)
R_F: shellfish toxin = 0.60; *G. catenella* toxin = 0.59.

In each case the chromatogram was cut into strips at every 0.5 cm, each strip was eluted with water and the solution injected into mice as a bioassay for toxic activity.

Hashimoto and Oshima[7] separated grammistins from soapfish (*Pogonoperca punctata*). Mucus from thawed fish can contain this toxin which shows hemolytic effects. The mucus was therefore scraped off the skin of a thawed fish and suspended in 100 ml of water. The aqueous solution was partitioned three times with an equal volume of 1-butanol. The butanol extracts were combined and evaporated to dryness under vacuum. After acidification with 5 ml of acetic acid, the residue was dissolved in 100 ml of distilled water and partitioned three times with an equal volume of diethyl ether to remove ether-soluble lipids. The aqueous layer was then partitioned with butanol to isolate the toxins.

Adsorbent: Silica Gel G
Solvent system 1: chloroform–methanol–water (65:25:4)
R_F: grammistin A = 0.28; grammistin B = 0.50; grammistin C = 0.07.
Solvent system 2: 1-butanol-1-propanol–0.1N hydrochloric acid (1:1:1)
R_F: grammistin A = 0.38; grammistin B = 0.49; grammistin C = 0.32.

Adsorbent: Avicel SF
Solvent A: 1-butanol–ethanol–water (4:1:2)
R_F: grammistin A = 0.58; grammistin B = 0.61; grammistin C = 0.46.
Solvent B: 2-butanol saturated with 1% acetic acid

R_F: grammistin A = 0.54; grammistin B = 0.61; grammistin C = 0.46.

The spots were detected with either ninhydrin or Dragendorff's reagent.

Goto[8] resolved several shellfish toxins by thin layer chromatography.

Adsorbent: Silica Gel G

Developing solvent: isopropanol–acetic acid–water (70:5:25)

R_F: maculotoxin = 0.52; tetrodotoxin = 0.53; saxtoxin = 0.53.

The spots were detected with a solution of ninhydrin and cyclohexamine.

A toxin which produced physiological effects similar to tetrodotoxin was extracted from a species of octopus by Croft and Howden.[9] The sample of octopus was extracted with methanol and filtered. The filtrate was chromatographed on a 20 g silica gel column. The column was eluted first with 100 ml of chloroform, then with 200 ml of methanol and chloroform (1:3), and finally with 100 ml of methanol and chloroform (1:1). The final eluate contained the octopus toxin, known as maculotoxin.

Adsorbent: Silica Gel G

Solvent: isopropanol–acetic acid–water (70:5:25)

R_F: maculotoxin = 0.1–0.2; tetrodotoxin = 0.6.

Adsorbent: Cellulose MN 300

Solvent A: 1-butanol–acetic acid–water (2:1:1)

R_F: maculotoxin = 0.2–0.3; tetrodotoxin = 0.55.

Solvent B: *t*-butanol–acetic acid–water (2:1:1)

R_F: maculotoxin = 0.3–0.4; tetrodotoxin = 0.65.

The spots were rendered visible with either ninhydrin or Dragendorff's reagent.

Neptunea arthritica, a marine gastropod, is widely eaten on the island of Hokkaido. It has sometimes caused poisoning with symptoms such as headache, dizziness, vomiting and urticaria with visual disturbances. Asano and Itoh[10] isolated the toxic compounds, using paper chromatography in conjunction with other procedures, and tested the toxicity of the isolated component on mice. Tetramine was identified as being the major constituent of the poison, which was contained in the salivary gland of the gastropod. Histamine and choline, which act as synergists, were also found.

Adsorbent: chromatography paper

Solvent 1: *n*-propanol–acetic acid–water (8:1:1)

Tetramine, R_F = 0.66 (the sample gave R_F values of 0.66 and 0.50).

Solvent 2: *n*-propanol–1N acetic acid–water (6:3:1)

Tetramine, R_F = 0.55 (the sample gave R_F values of 0.55 and 0.43).

Solvent 3: *n*-butanol–ethanol–water (7:2:1)

Tetramine, R_F = 0.47 (the sample gave R_F values of 0.47 and 0.35).

Spots were visualized with either Dragendorff or Chargaff-Levine reagent.[11]

6.4 PLANT ALKALOIDS

The ranges of the United States and other parts of the world support plants which contain metabolites potentially toxic to grazing cattle. In the past, thousands of deaths have occurred in areas that were highly infested with species of *Lupinus*, *Astragalus*, *Oxytropis*, *Delphinium* and other plants. These plants contain a range of alkaloids which may either be fatal to grazing animals or prove teratogenic to their offspring. Alkaloid chemicals are heterogeneous in structure, but they have in common the element nitrogen and a number of the characteristics of a base.

6.4.1 Colchicine alkaloids

These are found in all parts of the plant with the highest concentrations in the seeds and corms. They often occur in members of the *Liliaceae* family.

Some alkaloids of this type were chromatographed by Potesilova *et al.*[12]

Adsorbent: chromatographic paper impregnated with formamide
Developing solvent: benzene–chloroform–formamide (7:3:1)

R_F: colchicine = 0.26; cornigerine = 0.31; colchiceine = 0.20; demecolcine = 0.55; demecolceine = 0.31.

Adsorbent: Silica Gel G
Solvent 1: benzene–ethyl acetate–diethylamine (5:4:1)+85 ml methanol

R_F: colchicine = 0.56; cornigerine = 0.58; 3-demethylcolchicine = 0.32; 2-demethylcolchicine = 0.22; colchiceine = 0.27; isocolchicine = 0.38; lumicolchicine = 0.82.

Solvent 2: chloroform–acetone–diethylamine (7:2:1)

R_F: colchicine = 0.61; cornigerine = 0.63; 2-demethylcolchicine = 0.27; 3-demethylcolchicine = 0.36; colchiceine = 0.32; isocolchicine = 0.42; demecolcine = 0.70; demecolceine = 0.18; 2-demethyldemecolcine = 0.49; 3-demethyldemecolcine = 0.56.

These alkaloids fluoresce under long wavelength UV radiation. Some react with ferric chloride solution after treatment with hydrochloric acid vapors, others with Dragendorff's reagent. Some also develop selective colors when treated with sulfuric acid.

6.4.2 Ergot alkaloids

These alkaloids cause abortions in pregnant cattle because they produce the same effects as oestrogenic hormones. They are manufactured by the fungus *Claviceps purpurea* which parasitizes not only rye and wheat but the other grasses that grow in open pasture land.

Macek and Vanecek[13] applied paper chromatography to the resolution of some ergot alkaloids. A sheet of chromatographic paper was impregnated with 50% formamide and developed with chloroform. The R_F values obtained were: ergobasinine = 0.19; methylergobasinine = 0.33; ergotamine = 0.86. Macek[14] used the same immobile phase but developed the chromatogram with benzene and cyclohexane (1:1). The following R_F values were obtained: ergocristine = 0.28; ergocornine = 0.35; ergokryptine = 0.45; ergocristinine = 0.57; ergocorninine = 0.62; ergokryptinine = 0.72.

Fowler *et al.*[15] chromatographed some ergot alkaloids on silica gel.

Adsorbent: Merck Silica Gel F_{254} (pre-coated plates)

Solvent A: acetone

R_F: ergometrine maleate = 0.10; isolysergic acid = 0.11; methylergometrine = 0.13; methylsergide bimaleate = 0.13; dihydroergotamine = 0.19; lysergide = 0.27; ergotamine tartrate = 0.32; dihydroergocristine = 0.35; ergocristine = 0.51; ergocryptine = 0.51; ergosine = 0.31; ergocornine = 0.50.

Solvent B: acetone–chloroform (4:1)

R_F: D-lysergic acid = 0.03; ergometrine maleate = 0.44; isolysergic acid = 0.30; methylergometrine = 0.51; methylsergide bimaleate = 0.51; dihydro-ergotamine = 0.54; lysergide = 0.54; ergotamine tartrate = 0.63; dihydro-ergocristine = 0.64; ergocristine = 0.70; ergocryptine = 0.69; ergosine = 0.62; ergocronine = 0.68.

Solvent C: chloroform–methanol (4:1)

R_F: D-lysergic acid = 0.02; ergometrine maleate = 0.36; isolysergic acid = 0.32; methylergometrine = 0.42; methylsergide bimaleate = 0.54; dihydroergotamine = 0.60; lysergide = 0.60; ergotamine tartrate = 0.62; dihydroergocristine = 0.65; ergocristine = 0.68; ergocryptine = 0.67; ergosine = 0.61; ergocornine = 0.67.

Solvent D: chloroform–cyclohexane–isopropylamine (5:5:1)

R_F: ergometrine maleate = 0.07; isolysergic acid = 0.34; methylergometrine = 0.09; methylsergide bimaleate = 0.17; dihydroergotamine = 0.09; lysergide = 0.36; ergotamine tartrate = 0.11; dihydroergocristine = 0.19; ergocristine = 0.29; ergocryptine = 0.29; ergosine = 0.04; ergocornine = 0.17.

Pre-coated Merck Aluminum Oxide F_{254} Type E plates were also used for resolution of some ergot alkaloids.

Solvent 1: acetone

R_F: ergometrine maleate = 0.26; isolysergic acid = 0.66; methyl ergometrine = 0.31; methylsergide bimaleate = 0.33; dihydroergotamine = 0.40; lysergide = 0.70; ergotamine tartrate = 0.48; dihydroergocristine = 0.64; ergocristine = 0.66; ergocryptine = 0.68; ergosine = 0.52; ergocornine = 0.67; lysergamide = 0.27.

Solvent 2: 1,1,1-trichloromethane–methanol (9:1)

R_F: ergometrine maleate = 0.21; isolysergic acid = 0.59; methylergometrine = 0.24; methylsergide bimaleate = 0.45; dihydroergotamine = 0.54; lysergide = 0.61; ergotamine tartrate = 0.52; dihydroergocristine = 0.63; ergocristine = 0.66; ergocryptine = 0.65; ergosine = 0.49; ergocornine = 0.64.

The compounds were detected as blue fluorescent spots when the plate was sprayed with a solution of *p*-dimethylaminobenzaldehyde (1 g) in 10 ml of concentrated hydrochloric acid and 90 ml of ethanol. The plate required heating before being viewed in 254 or 250 nm radiation.

6.4.3 Tropane alkaloids

These compounds are found in the *Solanaceae* (potato family) and they have a blocking effect on the autonomic nervous system. Steinegger and Phokas[16] used paper chromatograms equilibrated in the ethyl acetate vapors and 25% formic acid for 14 h, developed in a solvent mixture of ethyl acetate and 25% formic acid (4:3). The R_F values obtained for the tropanes were: cuskhygrine = 0.19; scopoline = 0.30; tropine = 0.36; belladonine = 0.56; scopolamine = 0.59; hyoscyamine = 0.68; apoatropine = 0.80. The compounds were detected with Dragendorff's reagent.

Adamski *et al.*[17] coated plates with Silica Gel G suspended in an aqueous solution of 0.5N potassium hydroxide. The developing solvent was 80% methanol and ammonia (97:3). R_F values obtained were: tropine = 0.17; scopine = 0.32; scopolamine = 0.83; hyoscyamine = 0.51.

Oswald and Flück[18] also used silica gel.

Adsorbent: Silica Gel G

Solvent system: ethyl methyl ketone–methanol–7.5% ammonia (30:15:15)

R_F: tropine = 0.07; belladonine = 0.19; scopolamine = 0.67; hyoscyamine = 0.31; apoatropine = 0.41; aposcopolamine = 0.82; oscine = 0.50.

6.4.4 Quinolizidine alkaloids

Humans and cattle are sensitive to the quinolizidine alkaloids which occur in the *Fabaceae* and *Lythraceae* families. They produce gastrointestinal disturbances, irregular pulse coma and, in many cases, death. Jaminet[19] was able to isolate some of these alkaloids using paper chromatography. The chromatogram was developed in isobutanol, concentrated hydrochloric acid and water (100:15:27). The R_F values obtained were: *N*-methylcytisine = 0.18; cytisine = 0.21; anagyrine = 0.30; rhetamine = 0.32; hydroxylupanine = 0.35; lupinine = 0.47; genisteine = 0.49; pachycarpine = 0.50; sparteine = 0.52; lupinine = 0.76; spherocarpine = 0.95; orensine = 0.97; adenocarpine = 0.97.

Gill and Steinegger[20] used thin layer chromatography for their separations.

Adsorbent: Silica Gel G

Solvent system 1: cyclohexane–diethylamine (7:3)

R_F: N-methylcytisine = 0.30; cytisine = 0.07; anagyrine = 0.35; rhetamine = 0.80; hydroxylupinine = 0.19; hydroxysparteine = 0.54; lupanine = 0.61; sparteine = 0.93; isosparteine = 0.92; lupinine = 0.63; epilupinine = 0.56.

Solvent system 2: chloroform–methanol (4:1)

R_F: N-methylcytisine = 0.63; cytisine = 0.32; anagyrine = 0.60; rhetamine = 0.30; hydroxylupanine = 0.24; hydroxysparteine = 0.19; lupanine = 0.38; sparteine = 0.10; isosparteine = 0.03; lupinine = 0.12; epilupinine = 0.35; calycotomine = 0.46.

6.4.5 Ipecacuanha alkaloids

These alkaloids exhibit both emetic and cytotoxic properties; as a result some can be beneficial in small doses but extremely poisonous in large amounts. Auterhoff and Dietz[21] used paper chromatography for resolving them and a developing solvent of *n*-butanol, acetic acid and water (4:1:5). The following R_F values were obtained: rubremetine = 0.41; psychotrine = 0.45; *o*-methylpsychotrine = 0.47; cephaeline = 0.55; emetine = 0.58; emetamine = 0.65; corydaldine = 0.85.

Stahl[22] used thin layer chromatography with Silica Gel G as adsorbent and developed twice with a solvent mixture of chloroform and methanol (85:15). He obtained the following R_F values: psychotrine = 0.16; *o*-methylpsychotrine = 0.47; cephaeline = 0.13; emetine = 0.28; protoemetine = 0.63; emetamine = 0.67; 2-dehydroisoemetine = 0.18; 2-dehydroemetine = 0.21. Some of these compounds fluoresced under long-wavelength UV illumination; others were detected with a 0.5% iodine solution in chloroform. Some are also known to couple with diazonium salts to give colored compounds.

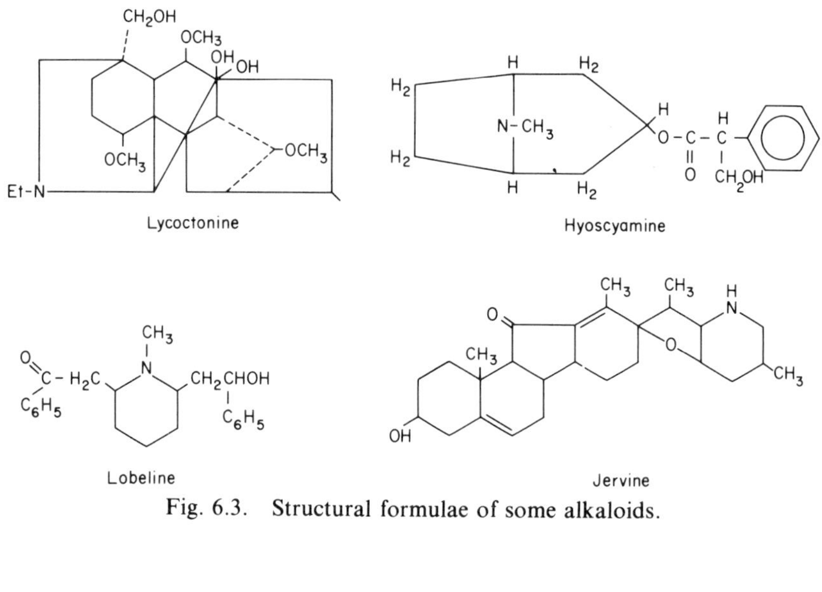

Fig. 6.3. Structural formulae of some alkaloids.

6.4.6 Lobelia alkaloids

This group of alkaloids are highly toxic to humans and animals, producing vomiting, paralysis and eventually death. Macek *et al.*[23] used paper chromatography to isolate lobeline. The paper was impregnated with a pH 3.8 citrate buffer and developed in a solvent mixture of *n*-butanol, water and citric acid (50:50:1). The R_F value for lobeline was 0.87. Waldi *et al.*[24] used instead thin layer chromatography on Alumina G adsorbent and developed in chloroform. The R_F value for lobeline in this case was 0.55. Dragendorff's reagent is commonly used to detect alkaloids.

Figure 6.3 shows the widely differing structures that can be found among the alkaloid compounds.

6.5 PLANT GLYCOSIDES AND GLUCOSIDES

6.5.1 Saponins

These are steroidal glycosides which are found in almost all varieties of plants but not always in quantities that are toxicologically significant. The saponin content of a plant varies with its species, its age and with the different organs of that plant which are chosen for consumption. Therefore, many plants which may carry foliage toxic to cattle also carry berries, fruit, flowers or seeds which may be perfectly safe for human consumption. Dust from saponin-containing plant material may produce crying, inflammation of the eyes, violent sneezing and a running nose. If, on the other hand, saponin-containing plants are consumed, salivation is increased, and nausea, emesis, stomach pain and gastroenteritis may ensue.

Boll[25] applied paper chromatography to the separation of some saponins. The chromatogram was developed with ethyl acetate, pyridine and water (3:1:3). The R_F values, relative to α-solanine as 1.00, were: tomatine = 0.85; polyanine = 1.11; soladulcamarine = 0.80; α-solamarine = 1.18; β-solamarine = 2.65; γ-solamarine = 3.80; solasonine = 0.88; solamargine = 2.92. When Silica Gel G was used as the thin layer adsorbent with a solvent composed of chloroform, ethanol and 1% ammonium hydroxide (2:2:1), the following R_F values were obtained: α-solanine = 0.26; α-chaconine = 0.36; tomatine = 0.29; soladulcamarine = 0.25; α-solamarine = 0.33; β-solamarine = 0.55; γ-solamarine = 0.78; solasonine = 0.33; solamargine = 0.55. Dragendorff's reagent and antimony trichloride rendered the spots visible.

Boll and Andersen[26] resolved other saponins on layers of Silica Gel G.

Solvent system A: chloroform–methanol (95:5)

R_F: tomata-3,5-diene = 0.75; diosgenin = 0.67; tigogenin = 0.67; yamogenin = 0.67; tomatidine = 0.47; tomatiden-3β-ol = 0.47; solasodine = 0.23; 5α-solasodan-3β-ol = 0.23.

Solvent system B: ethyl acetate–pyridine–water (45:15:45)

R_F: γ-solamarine = 0.52; solmargine = 0.44; β-solamarine = 0.38;

tomatine = 0.20; α-solamarine = 0.18; solasonine = 0.17; α-solanine = 0.22; α-chaconine = 0.35; soladulcamarine = 0.22; solamatgine = 0.45.

6.5.2 Cardiac glycosides

Plants such as the *Scrophulariaceae* contain cardiac glycosides (cardenolides) which affect the muscles of the heart. Some species of toads also secrete mixtures of cardiac glycosides called bufadienolides. Nover *et al.*[27] used paper chromatography in conjunction with immobile–mobile phase solvent systems to resolve many of these compounds. The paper was dipped into the immobile phase solution and then blotted between two pieces of paper. Schleicher and Schuell 2043 b chromatographic paper was used.

System A: immobile phase, 30% formamide in acetone; mobile phase, heptane–methyl ethyl ketone (3:2) saturated with formamide

R_F: digitoxigenin = 0.16; veneniferin = 0.32; bufalin = 0.25; uzarigenin = 0.10; oleandrin = 0.21; resibufogenin = 0.51.

System B: immobile phase, 35% formamide in acetone; mobile phase, methyl ethyl ketone–xylene (1:1) equilibrated with 3% formamide

R_F: digitoxigenin = 0.61; digitoxin = 0.46; nerifolin = 0.39; odorosid A = 0.67; menabegenin = 0.53; bufalin = 0.74; uzarigenin = 0.54; gitoxigenin = 0.62; gitosid = 0.78; gitoxin = 0.11; gitaloxigenin = 0.43; lanadoxin = 0.33; gitaloxin = 0.30; oleandrigenin = 0.51; oleandrin = 0.81; bufotalin = 0.59; digoxigenin = 0.09; sarmentogenin = 0.11; α-bufotalin = 0.17; telocinobufagin = 0.45; gomphosid = 0.30; acovenosigenin = 0.27; acovenosid A = 0.14; cannogenin = 0.34; apovannosid = 0.59; carpogenin = 0.26; strophanthidin = 0.50; cymarin = 0.42; hellebrigenin = 0.23; canarigenin = 0.53; xysmalogenin = 0.53; cerberigenin = 0.24; cerbertin = 0.67.

Thin layer plates coated with Silica Gel G were also used for separations.

Solvent A: *n*-hexane–ethyl acetate–ethanol (15:75:10) + 2.2% water

R_F: bufalin = 0.68; γ-bufotalin = 0.47; telocinobufagin = 0.49; 12β-hydroxybufalin = 0.45; desacetyl-bufotalin = 0.48; bufotalin = 0.56; bovogenin A = 0.46; hellebrigenin = 0.41; desglucohellebrin = 0.12; hellebrigenol = 0.33; resibufogenin = 0.67.

Solvent B: ethyl acetate–pyridine (90:10) + 2.7% water

R_F: bufalin = 0.75; telocinobufagin = 0.59; γ-bufotalin = 0.60; 12β-hydroxybufalin = 0.58; desacetylbufotalin = 0.60; hellebrigenin = 0.53; hellebrigenol = 0.46; scillarenin A = 0.71; scilliglaucosidin = 0.61; resibufogenin = 0.72.

Solvent C: methyl ethyl ketone + 5.4% water

R_F: bufalin = 0.76; telocinobufagin = 0.64; γ-bufotalin = 0.65; 12β-hydroxybufalin = 0.66; desacetylbufotalin = 0.65; hellebrigenin = 0.67; hellibrigenol = 0.51; scillarenin = 0.73; scilliglaucosidin = 0.76; resibufogenin = 0.76.

Solvent D: toluene–ethyl acetate–*n*–propanol (20:66:14) + 2.2% water

R_F: bufalin = 0.72; telocinobufagin = 0.51; γ-bufotalin = 0.51; 12β-hydroxybufalin = 0.57; desacetylbufotalin = 0.51; bufotalin = 0.65; hellebrigenin = 0.47; hellibrigenol = 0.33; scillarenin A = 0.71; resibufogenin = 0.70; scilliglaucosidin = 0.61.

6.5.3 Coumarin glycosides

Coumarin-type compounds, which are some of the most widely distributed phenols in the plant kingdom, are also known to have adverse physiological effects on animals. The simpler coumarins of some plants have already been identified as being toxic.

(i) *Esculetin* (6,7-dihydroxy coumarin) in the form of esculin 6-glucoside is the primary toxicant in a number of *Aesculus* species. Cattle and other animals may die from consuming leaves and nuts, while reports exist of children being killed by the nuts. Esculetin is a potent inhibitor of phenylalanine hydroxylase and can cause phenylketonuria in humans.

(ii) *Dicoumarol* (3,3′-methylenebis-4-hydroxycoumarin) is a poisonous phenol which is also from a plant source, and which has become useful as an inhibitor of blood clot formation. Cattle fed on sweet clover hay that is incompletely cured can die of a fatal haemorrhagic disease resulting from dicoumarol poisoning.

(iii) *Safrole* (1-allyl-3,4-methylenedioxybenzene) comprises about 80% of sassafras oil, and is also a component of star anise oil, camphor oil, nutmeg, mace and cinnamon leaf oils, as well as other essential oils. It is thought to be a carcinogen.

(iv) *Myristicin* (1-allyl-3,4-methylenedioxy-5-methoxybenzene) is an important component of nutmeg oil, nutmegs and mace. The consuming of large quantities of nutmeg can result in drowsiness, stupor or even death.

The above phenols and their glycosides may be extracted by homogenization with acetone, followed by filtration to remove suspended solid particles. The filtrate is partitioned with twice its volume of petroleum ether. If two phases do not form, water must be added until they do. The lower aqueous phase contains the polar phenols and their glycosides.

Stahl and Schorn[28] were able to resolve some natural coumarins by chromatography on Silica Gel G plates. The developing solvent mixture was toluene, ethyl formate and formic acid (50:40:10). The R_F values obtained were: aesculetin = 0.46; *nor*dalbergin = 0.58; umbelliferone = 0.59; marmesin = 0.56; dalbergin = 0.64; xanthotoxol = 0.63; psoralene = 0.69; bergapten = 0.68; xanthotoxin = 0.65; isopimpinellin = 0.67; *o*-methyldalbergin = 0.69; coumarin = 0.71; herniarin = 0.70; imperatorin = 0.73; osthol = 0.78.

Abdel-Hay *et al.*[29] also used Silica Gel G.

Solvent A: benzene–acetone (9:1)

R_F: umbelliferone = 0.27; marmesin = 0.18; dalbergin = 0.40; xanthotoxol = 0.22; psoralene = 0.75; bergapten = 0.75; xanthotoxin = 0.71; isopimpinellin = 0.70; *o*-methyldalbergin = 0.68; coumarin = 0.78; herniarin = 0.74; imperatorin = 0.79; osthol = 0.82.

Solvent B: benzene–ethyl acetate (9:1)

R_F: dalbergin = 0.15; psoralene = 0.47; bergapten = 0.42; xanthotoxin = 0.39; isopimpinellin = 0.36; *o*-methyldalbergin = 0.28; coumarin = 0.53; herniarin = 0.42; imperatorin = 0.49; osthol = 0.60.

Beyrich[30] prepared a thin layer plate by incorporating three parts of formamide and water (1:2) with the Silica Gel G and drying the mixture at room temperature for 18 h. Formamide constituted the immobile phase. The plate was then developed with dibutyl ether. The R_F values obtained were: bergapten = 0.31; xanthotoxin = 0.15; isopimpinellin = 0.20; coumarin = 0.22; imperatorin = 0.63; sphondin = 0.17; angelican = 0.32; isobergapten = 0.46; pimpinellin = 0.47; phellopterin = 0.70. The separated coumarins were detected by their flourescence under UV illumination at 365 nm or by spraying with an iodine–potassium iodide solution.

6.5.4 Goitrogenic glycosides

Glucosinolates are found mainly in cruciferous plants of the *Brassica* family. High concentrations of these compounds are found in the seeds of turnip, rutabaga, cabbage, broccoli, kale, kohlrabi, rape, brussel sprouts, and chinese cabbage. The ingestion of large amounts can cause hyperthyroidism (the enlargement of the thyroid gland due to inhibition of thyroid hormone formation). Goitrogenic compounds in cruciferous weeds can also enter cow's milk if dairy herds are allowed to graze on the particular crucifers mentioned above. In parts of the world where one particular goitrogenic food constitutes a major part of the diet, problems can arise, although the thorough cooking of such foods should inactivate the enzyme responsible for releasing the goitrogen.

These glycosides may be extracted with methanol. Wagner *et al.*[31] extracted 10 g of dried powdered seeds, or minced roots, with 50 ml of methanol, or 50 g of whole leaves or herbs with 250 ml of methanol. The solution was boiled, then filtered, and the solid material remaining was homogenized with 70%

methanol, and further boiled for 2 h. The solvent was filtered off and combined with the first filtrate. The combined extracts were evaporated down to about 5 ml and chromatographed on a cellulose powder MN100 column 9 mm in diameter and containing 5 g of the adsorbent. The glucosinolates were eluted from the column with methanol. The first 20 ml of eluate was discarded, the next 100 ml of eluate contained the glycosides. This eluate was evaporated down to a small volume for chromatography by either the paper or the thin layer method.

Schultz and Wagner[32] applied both paper and thin layer methods for resolving some of the glucosinolates. The paper techniques used were as follows.

Adsorbent: Schleicher and Schuell paper 2043b

Solvent 1: *n*-butanol–acetic acid–water (4:1:3) (descending development continuously for 70 h)

R_F (referenced to glucotropaeolin = 1): glucoiberin = 0.36; glucocheirolin = 0.39; sinigrin = 0.59; sinalbin = 0.67; glucolunarin = 0.67; glucocochlearin = 0.87; gluconasturtin = 0.25.

Solvent 2: *n*-butanol–pyridine–water (6:4:3) (descending continuous development for 45 h)

R_F (referenced to glucotropaeolin = 1): glucoiberin = 0.31; glucocheirolin = 0.51; sinigrin = 0.73; sinalbin = 0.93; glucolunarin = 0.79; glucocochlearin = 0.93; gluconasturtin = 1.15.

The migrated compounds were all detected by their reaction with alkaline silver nitrate when they formed violet spots against a yellow background.

The thin layer techniques of Schultz and Wagner[32] involved the use of Silica Gel G adsorbent with developing solvent consisting of *n*-butanol, *n*-propanol, acetic acid and water (3:1:1:1). The R_F values thus obtained were: glucoiberin = 0.16; glucocheirolin = 0.25; glucocapparin = 0.26; sinigrin = 0.37; sinalbin = 0.44; glucotropaeolin = 0.48.

The compounds were rendered visible by spraying with 25% trichloracetic acid solution, heating for 10 min at 140 °C and then respraying with a 5% ferric chloride and 1% potassium ferricyanide solution (1:1). The glucosinolates reacted to form blue spots on a yellow background.

Further positive identification of the glucosinolates could be made by forming the corresponding thiourea from the conjugate. This was done by hydrolyzing the methanol extract at pH 7 with tyrosinase and a few milliliters of ascorbic acid solution. After hydrolyzing for 2 h, the mustard oils were extracted with ether, and 1 ml of ethanol and 1 ml of concentrated ammonium hydroxide were added. The mixture was heated for 20 min in a water bath to form the thioureas.

The thioureas formed as a result of this procedure are: methyl thiourea from glucocapparin; ethylthiourea from glucolepidin; isopropyl thiourea from glucoputranivin; allyl thiourea from sinigrin; *a-sec*-butyl thiourea from glucocochlearin; *β*-phenylethyl thiourea from gluconasturtin; benzyl and phenyl thiourea from glucotropaeolin; methylsulfinyl propyl thiourea from glucoiberin; and methylsulfonyl propyl thiourea from glucocheirolin.

Schultz and Wagner[32] used Silica Gel G thin layer adsorbent in conjunction with a solvent system of ethyl acetate, chloroform and water (3:3:4). This solvent mixture formed a two-phase system; the upper phase was used for development. The R_F values obtained were: methyl thiourea = 0.20; ethyl thiourea = 0.30; isopropyl thiourea = 0.41; allyl thiourea = 0.43; α-*sec*-butyl thiourea = 0.49; β-phenylethyl thiourea = 0.51; benzyl thiourea = 0.57; phenyl thiourea = 0.67.

The sulfoxide methylsulfinyl propyl thiourea and sulfone methylsulfonyl propyl thiourea had to be separated by paper chromatography.[31]

Adsorbent: Schleicher and Schuell 2043b chromatographic paper
Solvent: *n*-butanol–acetic acid–water (4:1:3) (descending development)

R_F: methylsulfinyl thiourea = 0.54; methylsulfonyl propyl thiourea = 0.57.

Thin layer chromatography using a fluorescent indicator as solvent distinguished the thioureas as quenched spots. Ferric chloride–ferricyanide reagent could also be used for detection.

6.5.5 Cyanogenetic glucosides

Cyanide in trace amounts is an almost universal constituent of plants. It occurs mainly as a cyanogenetic glucoside. Relatively high concentrations are found in some grasses, legumes, root crops and fruit kernels. Ingestion of large amounts of these glucosides can cause generalized muscular paresis, mental confusion and respiratory impairment, which are the usual symptoms of acute cyanide poisoning. In the tropics, cyanide poisoning is also caused by eating cassava that has been boiled for excessively long periods.

The cyanogenetic glucosides can be extracted from the green parts of the plant or from the powdered seed by refluxing with boiling ethanol, which is then evaporated and the residue is extracted with water. The water solution is centrifuged and the supernatant liquid is evaporated to dryness, leaving a syrupy residue. This syrup is extracted with methanol and filtered. The methanol filtrate is then evaporated and the residue is redissolved in methanol. Tantisewie *et al.*[33] used paper chromatography for resolution of some of the glucosides.

Adsorbent: Whatman No. 1 paper or Schleicher and Schuell 2043b
Solvent 1: methyl ethyl ketone–acetone–water (15:5:3)

R_F: amygdalin = 0.35; prunasin = 0.90; linamarin = 0.54; lotaustralin = 0.67; gynocardin = 0.41.

Solvent 2: *n*-propanol–water (7:3)

R_F: amygdalin = 0.72; prunasin = 0.85; linamarin = 0.77; lotaustralin = 0.83; gynocardin = 0.66.

Solvent 3: ethyl acetate–acetone–water (4:5:1) (equilibrate for 1 h)
R_F: amygdalin = 0.37; prunasin = 0.86; linamarin = 0.54; lotaustralin = 0.66; gynocardin = 0.41.

Solvent 4: *n*-butanol–ethanol–water (7:2:2)
R_F: amygdalin = 0.43; prunasin = 0.69; linamarin = 0.55; lotaustralin = 0.66; gynocardin = 0.42.

The chromatogram was sprayed with a linamarase–emulsin solution to release the hydrocyanic acid and glucose. The glucose which was released was detected by spraying with a benzidine–trichloracetic acid solution and heating.

6.6 SELENIUM COMPOUNDS

Cabbage (*Brassica oleracea capitata*) is a vegetable commonly grown in family gardens. It is capable of absorbing, metabolizing and storing large quantities of selenium in its tissues. Thus there is a possibility that the consumption of cabbages containing high levels of selenium could result in selenium poisoning. Hamilton[34] was able to separate some selenium analogs of sulfur-containing amino acids with thin layer chromatography.

Adsorbent: Silica Gel H (150 mm thick)

Solvent 1: 2-methyl-1-propanol–acetic acid–chloroform–water–methanol (40:15:10:15:20)

The chromatograms were developed twice and the resulting R_F values were divided by 150. The relative separation values were: selenopeptides and proteins = 5; Se-methylselenomethionine = 11; selenocystathionine = 0.31; selenocystine = 45; Se-methylselenocysteine selenoxide = 54; selenohomocystine = 62; selenocysteine type compound = 87; Se-methylselenocysteine = 95; selenomethionine = 109.

Solvent 2: l-propanol–ammonium hydroxide (70:30)

Separation values given by this system were: selenopeptides and proteins = 3; Se-methylselenomethionine = 14; selenocystathionine = 32; selenocystine = 40; Se-methylselenocysteine selenoxide = 55; selenohomocystine = 48; selenocysteine-type compound = 64; Se-methylselenocysteine = 87; selenomethionine = 96.

Solvent 3: l-butanol–acetic acid–water (60:15:25)

Separation values given by this system were: selenopeptides and proteins = 3; Se-methylselenomethionine = 9; selenocystathionine = 17; selenocystine = 34; Se-methylselenocysteine selenoxide = 26; selenohomocystine = 48; selenocysteine-type compound = 58; Se-methylselenocysteine = 67; selenomethionine = 85.

Ninhydrin, naphthoresorcinol or sulfuric acid were used as detecting reagents; where labelled compounds were used, radioautography was employed for detection.

6.7 MYCOTOXINS

Mycotoxins have been recognized as important natural poisons. They are molecules which are natural products of the metabolism of some fungi. Although most fungi can grow on almost any type of decaying organic matter, the species which generally produce mycotoxins are parasites which attack living tissue and cause its decomposition.

Reports of mycotoxicosis are available throughout recorded history. In mediaeval times, a number of outbreaks of ergotism, a condition caused by the fungus *Claviceps purpurea*, were recorded. At the end of World War II thousands of Russians developed toxic alimentary Alukia by eating grain that had overwintered in the fields and was mouldy. The condition was attributed to fusariogenin from *Fusarium poae* and to the epi- and fagicladosporic acids produced by the species *Cladosporium*.

Also, during World War II, many Japanese began to show symptoms similar to those of beri-beri. It was later discovered to be yellow rice toxicosis caused by ingestion of the mycotoxins islanditoxin, patulin, luteoskyrin, citrinin, citreoviridin and others, that were produced by species of *Aspergillus* and *Penicillium* growing on rice.

Figures 6.4 and 6.5 show the structural formulae of some mycotoxins. Their toxic effects are as variable as their composition. They can be hepatoxic, nephrotoxic, neurotoxic, hemmorhagic, dermatoxic, genitotoxic, teratogenic or carcinogenic.

The discovery by Lancaster *et al.*[35] and Sargeant *et al.*[36] of serious aflatoxicosis in farm animals and of the presence of aflatoxins in their feed, triggered a worldwide effort to develop analytical methods for these and other

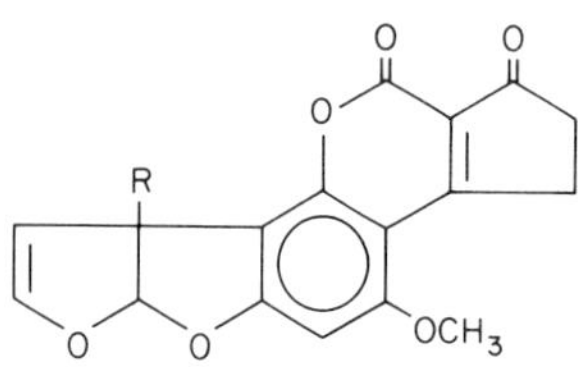

R=H Aflatoxin B_1
R=OH Aflatoxin M_1

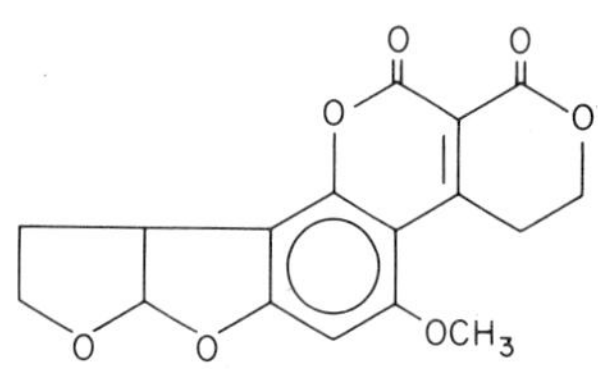

Aflatoxin G_1

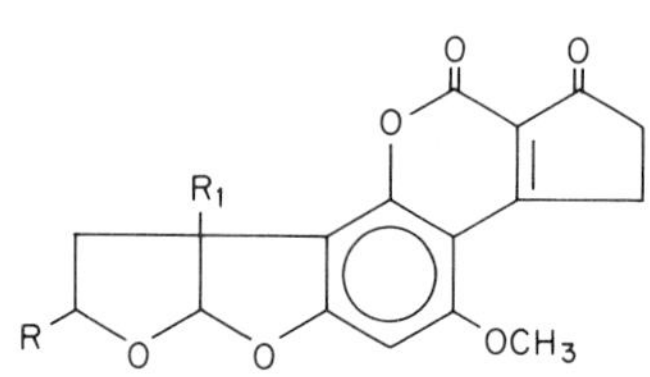

R = H, R_1 = H Aflatoxin B_2
R = H, R_1 = OH Aflatoxin M_2

Fig. 6.4. Structural formulae of mycotoxins.

mycotoxins. The work of Sargeant's group involved the extraction of the toxic component from Brazilian ground-nut meal which had previously caused the death of about 100 000 turkeys. The meal was subjected to Soxhlet extraction with methanol and the toxic component was isolated by paper chromatography. The chromatogram was developed with 5% acetic acid in *n*-butanol. A fluorescent spot was located at $R_F = 0.7$, which was eluted and found to contain the toxic principle. When thin layer chromatography was applied to the analysis of this toxic principle, it became resolved into several components. Thin layer chromatography was therefore preferred for such analyses because of its simplicity and sensitivity.

Nesbitt *et al.*[37] used thin layer chromatography on the toxic principle already extracted by paper chromatography. Alumina plates were used with chloroform and methanol (98.5 + 1.5) as the developing solvent. Two spots were resolved, one with an R_F value of 0.6 and another with a slightly lower R_F. The lower spot fluoresced green and the upper spot blue-violet.

Many food products, including peanuts, contain aflatoxins. This fact caused great concern to the United States Food and Drug Administration because of the widespread consumption of peanut butter, especially by children Nesheim *et al.*[38] introduced a mechanical homogenization procedure using methanol as the extracting solvent. Fifty grams of peanut product was homogenized with 100 ml of hexane and 250 ml of methanol and water (55:45). The homogenate was centrifuged to produce two layers, a top layer of hexane and a bottom layer of aqueous methanol. The hexane was included in order to remove some of the lipophilic interferences.

Fifty milliliters of the bottom layer were pipetted into a beaker containing 55 g of acid-washed celite, which was sufficient to soak up all the sample.

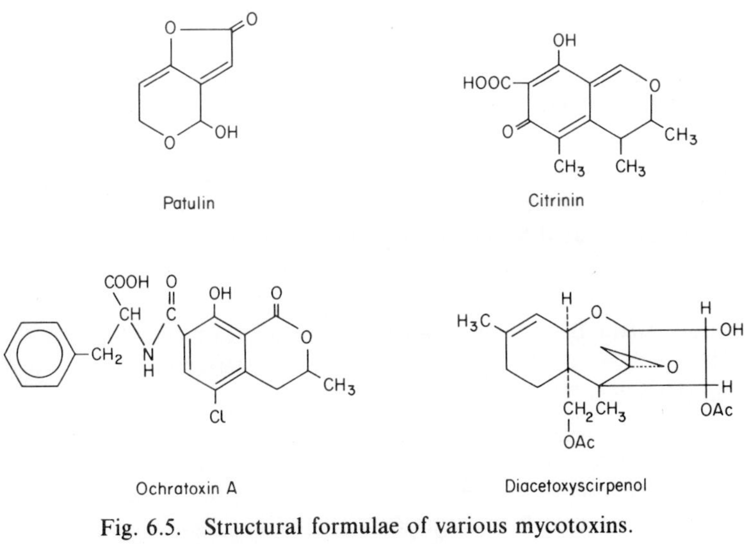

Fig. 6.5. Structural formulae of various mycotoxins.

The celite was then packed into a 45 mm ID chromatographic column. The column was first washed with 400 ml of hexane. The empty beaker was then washed out with 70 ml of hexane and chloroform (50:50) and the washings transferred to the column. The column was then eluted with 430 ml of hexane–chloroform mixture, and finally with 200 ml of chloroform. All the fractions were transferred to 50 ml Erlenmeyer flasks and evaporated to dryness on a steam bath. Each sample was rinsed with 500 μl of chloroform and then spotted on thin layer plates.

Many procedures have been devised for extracting and isolating mycotoxins from food products. A recent one was developed by Roberts and Patterson.[39] Twenty five grams of finely-ground sample were weighed into an 8 oz wide-mouthed bottle. Into the bottle was added 100 ml of acetonitrile and 4% potassium chloride (9:1). The bottle was closed with a screw cap and aluminum liner. The toxin was extracted by a wrist-action shaker for 30 min. The resulting solution was filtered and 50 ml of the filtrate was partitioned with 50 ml of 2,2,4-trimethylpentane. The upper hydrocarbon layer was discarded and the lower acetonitrile layer was partitioned again with 50 ml of 2,2,4-trimethylpentane. The acetonitrile layer had 12.5 ml of distilled water added to it and was then extracted with 25 ml of chloroform. The chloroform layer was drawn off, and three further extractions were made with 10 ml volumes of chloroform. All the partitioned volumes were filtered through anhydrous sodium sulfate, and the combined filtrates were labelled 'extract 1'.

The aqueous layer remaining in the separatory funnel was made acid by the addition of 1 ml of 1.0N hydrochloric acid. It was then extracted with three 10 ml volumes of chloroform. These extracts were filtered through sodium sulfate, combined, and labelled 'extract 2'.

Both extracts were evaporated down to 0.1 ml volume for thin layer chromatography. If further clean-up is necessary for extract 1, it may be re-dissolved in 1.25 ml of acetonitrile for a membrane clean-up. A 20 cm 8/32 Visking dialysis tube was washed in distilled water, and 1 ml of the acetonitrile solution was transferred to the sac. It was equilibrated with 25 ml of 30% acetone for 16 h with constant gentle shaking. The aqueous acetone dialysate was extracted with three 10 ml volumes of chloroform, filtered through sodium sulfate and evaporated to a small volume for chromatography.

Adsorbent: precoated MN Silical Gel HR plates

Solvent system 1: toluene–ethyl acetate (1:3)

R_F: aflatoxin B1 = 0.33; aflatoxin G1 = 0.25; diacetoxyscirpenol = 0.48; patulin = 0.53; penitrem A = 0.75; sterigmatocystin = 0.75; zearalenone = 0.72.

Solvent system 2: toluene–ethyl acetate–acetone (1:4:4)

R_F: aflatoxin B1 = 0.16; aflatoxin G1 = 0.07; citrinin = 0.98; diacetoxyscirpenol = 0.19; patulin = 0.26; penitrem A = 0.58; sterigmatocystin = 0.70; zearalenone = 0.64.

Solvent system 3: chloroform–methanol–acetone (1:1:5)
R_F: aflatoxin B1 = 0.88; aflatoxin G1 = 0.88; citrinin = 0.94; diacetoxyscirpenol = 0.92; ochratoxin A = 0.10, 0.20; patulin = 0.57; penitrem A = 0.64; sterigmatocystin = 0.92; zearalenone = 0.93.

Solvent system 4: toluene–ethyl acetate–formic acid (6:3:1)
R_F: aflatoxin B1 = 0.32; aflatoxin G1 = 0.20; citrinin = 0.52; diacetoxyscirpenol = 0.37; ochratoxin A = 0.62; patulin = 0.45; penitrem A = 0.69; sterigmatocystin = 0.68; zearalenone = 0.66.

Solvent system 5: acetone–chloroform (7:93)
R_F: aflatoxin B1 = 0.40; aflatoxin G1 = 0.29; diacetoxyscirpenol = 0.28; patulin = 0.22; penitrem A = 0.35; sterigmatocystin = 0.72; zearalenone = 0.48.

Solvent system 6: chloroform–ethyl ether–acetic acid (85:15:5)
R_F: aflatoxin B1 = 0.56; aflatoxin G1 = 0.42; diacetoxyscirpenol = 0.52; citrinin = 0.72; ochratoxin A = 0.68; patulin = 0.42; penitrem A = 0.68; sterigmatocystin = 0.87; zearalenone = 0.71.

Development was carried out after equilibration and the compounds were detected by their fluorescence in UV radiation. Ferric chloride was used to detect penitrem A as a blue-green spot.

Steyn[40] impregnated a Silica Gel G layer with oxalic acid and developed with chloroform and methyl isobutyl ketone (4:1). The following R_F values were found: aspertoxin = 0.12; ochratoxin B = 0.20; secalonic acid D = 0.23; aflatoxin G1 = 0.30; aflatoxin B1 = 0.40; ochratoxin A = 0.48; cyclopiazonic acid = 0.65; zearalenone = 0.72; sterigmatocystin = 0.85.

Durackova *et al.*[41] used Silica Gel G and precoated (Silufol) plates for resolving the aflatoxins.

Solvent system A: benzene–methanol–acetic acid (24:2:1)
R_F: aflatoxin B1 = 0.28; aflatoxin B2 = 0.30; aflatoxin G1 = 0.19; aflatoxin G2 = 0.23; aflatoxin M1 = 0.22. R_F values obtained with Silufol were: aflatoxin B1 = 0.28; aflatoxin B2 = 0.29; aflatoxin G1 = 0.26; aflatoxin G2 = 0.25.

Solvent system B: toluene–ethyl acetate–90% formic acid (6:3:1)
R_F: aflatoxin B1 = 0.32; aflatoxin B2 = 0.18; aflatoxin G1 = 0.19; aflatoxin G2 = 0.08; aflatoxin M1 = 0.16. R_F values obtained with Silufol were: aflatoxin B1 = 0.21; aflatoxin B2 = 0.16; aflatoxin G1 = 0.16; aflatoxin G2 = 0.11.

Solvent system C: benzene–ethanol (95:5)
R_F: aflatoxin B1 = 0.20; aflatoxin B2 = 0.14; aflatoxin G1 = 0.14; aflatoxin G2 = 0.08; aflatoxin M1 = 0.1. R_F values obtained with Silufol were: aflatoxin B1 = 0.13; aflatoxin B2 = 0.13; aflatoxin G1 = 0.11; aflatoxin G2 = 0.09.

Solvent system D: chloroform–acetone (9:1)
R_F: aflatoxin B1 = 0.33; aflatoxin B2 = 0.33; aflatoxin G1 = 0.25;

aflatoxin G2 = 0.17. R_F values obtained with Silufol were: aflatoxin B1 = 0.32; aflatoxin B2 = 0.22; aflatoxin G1 = 0.22; aflatoxin G2 = 0.19. Solvent system A used with Silica Gel G gave the following R_F values for some other mycotoxins: ochratoxin A = 0.52; ochratoxin B = 0.41; ochratoxin C = 0.80; patulin = 0.34. The same solvent system used with Silufol gave the following R_F values: citrinin = 0.21; ochratoxin A = 0.29; patulin = 0.24.

Josefsson and Möller[42] developed a screening method for some mycotoxins that may occur in cereals. A cereal sample of 50 g was placed in a 500 ml round-bottomed flask. A 25 ml volume of 0.1M phosphoric acid and 250 ml of chloroform were added to the flask, and the sample was extracted for 30 min with a wrist-action shaker.

The extract was filtered through 10 g of celite on a glass-fiber filter fitted into a Buchner funnel. A 25 ml aliquot of the filtrate was taken and evaporated to a volume of which about 5 ml was transferred to a small glass tube and further evaporated to 0.25 ml by heating in a nitrogen atmosphere. Corn meal samples had to be extracted by mechanical homogenization with the same solvent.

The sample was cleaned up with a gel filtration column. Twenty five grams of Sephadex LH-20 was allowed to absorb a solution of chloroform and methanol (2:1) for about 30 min. The excess solvent was poured off and the slurry added to a chromatographic column (15 mm ID × 600 mm) until the gel had settled to a height of 59 cm. The column was washed with the chloroform–methanol solution with the aid of a peristaltic pump, at a rate of 100 ml h^{-1} for 2 h.

The sample was added to the column and eluted with solvent mixture at a rate of 75 ml h^{-1}. Aflatoxin B1 was eluted by the first 56–64 ml, aflatoxins B2, G1 and G2 were eluted by the first 55–64 ml, ochratoxin A was eluted by the first 59–64 ml, patulin was eluted by the first 78–88 ml, sterigmatocystin was eluted by the first 61–66 ml, and zearalenone was eluted by the first 64–68 ml. Therefore, the eluent fraction between 53 and 75 ml contained the aflatoxins, ochratoxin A, sterigmatocystin, and zearalenone, while the eluent fraction collected between 76 and 90 ml contained patulin.

The two fractions were evaporated down to 5 ml, transferred to a small tube and evaporated just to dryness in a nitrogen atmosphere. Each sample was dissolved in 100 μl of chloroform for application to the thin layer plate.

Adsorbent: Silica Gel G-HR

Solvent 1: benzene–hexane (3:1)

This solvent system was used to separate the lipids from the mycotoxins. The first 50 minutes of development moved the lipids with the solvent though the mycotoxins remained at the origin. After this first development, the plate was removed from the tank, air-dried, and developed in solvent 2, a mixture of benzene, ethyl acetate and formic acid (80:20:0.5). The following R_F values were obtained: aflatoxin B1 = 0.06; aflatoxin B2 = 0.11; aflatoxin G1 = 0.04;

aflatoxin G2 = 0.06; ochratoxin A = 0.27; patulin = 0.30; sterigmatocystin = 0.72; zearalenone = 0.63. A third solvent system of chloroform and acetone (9:1) gave the following R_F values: aflatoxin B1 = 0.54; aflatoxin B2 = 0.61; aflatoxin G1 = 0.41; aflatoxin G2 = 0.45; patulin = 0.44.

The compounds were detected under UV illumination. Smaller amounts of zearalenone and sterigmatocystin could be detected by spraying the plate with a 20% solution of aluminum trichloride and heated in an oven at 100 °C for 5 min. Patulin was detected by spraying the plate with a 0.5% aqueous solution of 3-methylbenzo-2-thiazalone hydrazone hydrochloride, and heated at 120 °C for 15 min. It was then visible as a yellow spot.

REFERENCES

1. P. M. Scott, B. B. Caldwell and G. S. Wiberg, *Food Cosmet. Toxicol.* **9**, 179 (1971).
2. V. Palyza, *J. Chromatogr.* **64**, 317 (1972).
3. G. Sullivan, L. R. Brady and V. E. Tyker Jr, *J. Pharm. Sci.* **54**, 921 (1965).
4. N. Trieff, J. J. Spikes, S. M. Ray and J. B. Nash in *Toxins of Animal and Plant Origin*, Ed. by de Vries and Kochva. Vol. 2, p. 557. Gordon and Breach, New York (1972).
5. Y. Oshima, W. E. Fallon, Y. Shimizu, T. Noguchi and Y. Hashimoto, *Jpn Soc. Sci. Fish.* **42**, 851 (1976).
6. J. M. Burke, J. Marchisotto, J. J. A. McLaughlin and L. Provaoli, *N.Y. Acad. Sci. Ann.* **60**, 837 (1960).
7. Y. Hashimoto and Y. Oshima, *Toxicon* **10**, 279 (1972).
8. Y. Goto, *Tetrahedron* **21**, 2059 (1965).
9. J. A. Croft and M. E. H. Howden, *Toxicon* **10**, 645 (1972).
10. M. Asano and M. Itoh, *N.Y. Acad. Sci. Ann.* **60**, 674 (1960).
11. E. Chargaff, C. Levine and C. Green, *J. Biol. Chem.* **175**, 67 (1948).
12. N. Potesilova, J. Hrbek Jr and F. Santavy, *Collect. Czech. Chem. Commun.* **32**, 141 (1967).
13. K. Macek and S. Vanacek, *Pharmazie* **10**, 422 (1955).
14. K. Macek in *Chromatography*, Ed. by E. Heftman. Rheinhold, New York, 1967, p. 606.
15. R. Fowler, P. J. Gomm and D. A. Patterson, *J. Chromatogr.* **72**, 351 (1972).
16. E. Steinegger and G. Phokas, *Pharm. Acta Helv.* **31**, 284, 330 (1956).
17. R. Adamski, J. Lutomski and J. Wisniewski, *J. Deut. Apothkr Zeit.* **107**, 185 (1967).
18. N. Oswald and H. Flück, *Pharm. Acta Helv.* **39**, 293 (1964).
19. F. Jaminet, *J. Pharm. Belg.* **36**, 9 (1954).
20. S. Gill and E. Steinegger, *Pharm. Acta Helv.* **39**, 508, 565 (1964).
21. H. Auterhoff and H. Dietz, *Arch. Pharm.* **289**, 422 (1956).
22. E. Stahl in *Thin Layer Chromatography*, Ed. by E. Stahl. Springer Verlag, New York, 1969, p. 445.
23. K. Macek, J. Vercerkava and J. Stanislavova, *Pharmazie* **20**, 605 (1965).
24. D. Waldi, K. Schnackerz and F. Munter, *J. Chromatogr.* **6**, 61 (1961).
25. P. M. Boll, *Acta Chem. Scand.* **16**, 1819 (1962).
26. P. M. Boll and B. Andersen, *Planta Med.* **10**, 421 (1962).
27. L. Nover, G. Jüttner, S. Noacks, G. Baumgarten and M. Leuckner, *J. Chromatogr.* **39**, 419 (1969).
28. E. Stahl and P. J. Schorn, *Z. Physiol. Chem.* **325**, 263 (1961).

29. F. M. Abdel-Hay, E. A. Abu-Mustafa, B. A. H. El-Tamil and M. B. E. Fayez, *Planta Med.* **13**, 91 (1965).
30. T. Beyrich, *J. Chromatogr.* **20**, 173 (1965).
31. H. Wagner, L. Hörhammer and H. Nufer, *Arzneim. Forsch.* **15**, 453 (1965).
32. O. E. Schultz and W. Wagner, *Z. Naturforsch.* **11b**, 73 (1956).
33. B. Tantisewie, H. W. L. Ruijgrok and R. Hegnauer, *Pharm. Weekblad.* **104**, 1341 (1969).
34. J. W. Hamilton, *J. Agric. Food Chem.* **23**, 1150 (1975).
35. M. C. Lancaster, F. P. Jenkins and J. McL. Phelp, *Nature* **192**, 1095 (1961).
36. K. Sargeant, A. Sheridan, J. O'Kelly and R. B. A. Carnaghan, *Nature* **192**, 1096 (1961).
37. B. F. Nesbitt, J. O'Kelly, K. Sargeant and A. Sheridan, *Nature* **195**, 1062 (1962).
38. S. Nesheim, D. Banes, L. Stoloff and A. D. Campbell, *J. Assoc. Off. Agric. Chem.* **47**, 586 (1964).
39. B. A. Roberts and D. S. P. Patterson, *J. Assoc. Off. Anal. Chem.* **58**, 1178 (1975).
40. A. Steyn, *J. Chromatogr.* **45**, 473 (1969).
41. Z. Durackova, V. Betina and P. Nemel, *J. Chromatogr.* **116**, 141 (1976).
42. B. G. Egon. Josefsson and T. E. Möller, *J. Assoc. Off. Anal. Chem.* **60**, 1369 (1977).

CHAPTER 7

CHROMATOGRAPHIC TECHNIQUES AND REAGENTS

7.1 INTRODUCTION

Paper chromatography has been an extremely useful technique for separating organic and inorganic compounds in the past, but is being rapidly supplanted by thin layer chromatography. However, many of the paper techniques can be altered to suit thin layers of cellulose in its many forms. This discussion will be restricted to observations made by the author when applying thin layer chromatography to pesticide residue determinations, and to information from other investigators which may prove helpful in developing a satisfactory method for any particular problem.

Block *et al.*,[1] Hais and Macek,[2] and Smith[3] have written books on paper chromatography which are recommended reading for those who wish to learn more about the subject.

7.2 ADSORBENTS

Any material can be used as an adsorbent provided it forms an interconnected system of pores or capillaries which allow solvents to ascend or descend through them. The main adsorbents used for thin layer chromatography are silica gel, alumina, and cellulose, though Polyamide has recently been applied as a thin layer medium for the separation of molecules with hydrogen bonding properties.

Silica gel appears to be the adsorbent most widely used, since the number of publications citing its use far exceed those for other media. It is preferred for several reasons: it is relatively unreactive towards bioactive chemicals, it undergoes little irreversible adsorption or decomposition, it can be prepared in a very pure form, and its pore size can be controlled.

As far as is known to the author there are only three world producers of chromatographic silica gel; two in Europe and one in the United States. The main differences in manufacture derive from the type of acid used for precipitating the gel, the duration of contact between the gel and the acid, which determines pore size, and the degree of purity of the final product. The average size and distribution of the pores affects the separating ability of the gel. Small pores result in adsorbents of high surface area and larger pores produce adsorbents of lower surface area.

A small-pore gel will give a lower R_F value for a polar compound than a large pore gel, when the same developing solvents are used. The first silica gels used had average pore sizes of 4 nm or less. Today, silica gels produced in Europe have an average pore size of 6 nm while the American silica gels are manufactured in 4 nm and 8 nm forms.

An analyst must know the specifications of the silica gels used if he intends to reproduce a previously published procedure. If the pore size available for an experiment differs greatly from those reported as used by other workers, the analyst must recognize that his results will differ from those reported, and that an adjustment may be required in the composition of the developing solvent. Generally, higher R_F values will be obtained with gels of larger pore size.

Recently, a thin layer alumina with pore size designation has become available. A description of pore size can be helpful to the analyst, especially when he purchases differently manufactured batches. This appears to be the trend with alumina manufacturers. It would be helpful if alumina could also be designated by its crystalline form since several different types exist.

Cellulose is obtainable as fibers, microcrystalline and also chemically altered forms, e.g., acetylated, and with ion-exchange groups added on. The advantage of using the microcrystalline form is that it produces more compact spots. This helps in the resolution of compounds. Fibrous cellulose gives a diffusion pattern similar to that in paper chromatograms.

In 1968 Merck introduced precoated plates which were given a mixed reception. They have proved to be an excellent idea, however, and today a wide range of high quality precoated plates are available. The major differences between these plates derive from the binders used and from the support used for the thin layer.

Calcium sulfate and starch were the original binders used, when thin layer plates were coated in the laboratory, with window glass as the support. Commercially produced plates contain binders composed of inorganic and organic polymers. They may have supports of glass, aluminum or plastic. The glass used by commercial coaters is quite different from that used in the laboratory, being more porous and receptive to electrostatic binding.

The plates may be obtained as microslides, as 20×20 cm plates, and in many intermediate sizes. Some companies will coat plates with special adsorbents at the request of the purchaser. The 'hard layer' plates have excellent abrasion resistance and many can even be dipped into an aqueous solution without the adsorbent flaking off the support.

The author has been using precoated plates for almost 10 years and has found those commercially available today to be of excellent quality. The only difficulty recently has been the introduction of new binders without notification to potential users. Binders can interfere with chromogenic reagents. A problem of this nature arose when *p*-nitrobenzyl pyridine was used as a detecting agent for organophosphate pesticides. It produced excellent results until a new batch of precoated plates was used, when no positive results were obtained. A new binder had been introduced which conferred acid characteristics to the plate and, since alkaline conditions were necessary for the reagent to work, it inactivated the entire detection system.

The pH of the adsorbent layer can be measured by scraping off a 2 cm^2 area, shaking it with 10 ml of distilled water for several minutes and then measuring its pH with a pH meter. This is probably a good procedure to follow with every new batch of plates purchased.

The aluminum backed plates open up a new approach to reagents for visualizing chromatographed compounds. The value to chromatography of many of the Feigl spot tests and other tests requiring high temperature can now be assessed. These reactions will be discussed in detail in Section 7.7.

7.3 SAMPLE APPLICATION

Samples and standards should be applied to the plate before development. This is accomplished with the use of various micropipettes or syringes. Figure 7.1 shows examples of applicators (spotting devices) which may be used. Figure

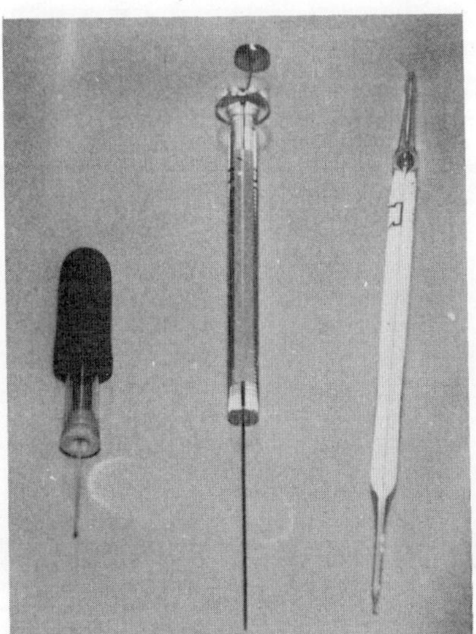

Fig. 7.1. Some examples of applicators or spotting devices for thin layer plates.

7.1(a) shows a calibrated 2 μl capillary which is accurate and may be used for quantitative work, sizes from 1 to 10 μl are available; (b) shows a gas chromatographic syringe (10 μl) which is useful for qualitative work but is not accurate enough for quantification; (c) shows a commercial 1 μl pipette which gives reproducibile results.

When environmental samples are known to contain high concentrations of the compound to be determined, microvolumes (μl aliquots) of standards and samples may be used for spotting on thin layer plates. In many cases, however, the total isolated sample must be used. This usually means that a large volume of solvent must be reduced to a few microliters before application to the plate. Let us consider an example where the volume of eluate containing the compound to be identified is 200 ml. This can be evaporated to a small volume or to dryness by blowing air or an inert gas over the surface of the solvent contained in a flask or beaker. The solvent can also be evaporated in a rotary evaporator. However, the containers used for either procedure do not facilitate the transfer of the solution to a chromatogram.

Instead, it is more efficient to dissolve the residue in 10–25 ml of solvent, transfer the solution to a conical tube and evaporate once again just to dryness or to 50–100 μl. If the contents contain no solvent, 50 μl of suitable solvent is added to the tube, rinsing down the sides. This rinsing procedure is repeated several times. The sample is then ready for transfer to the plate. This technique is suitable for qualitative work but not for quantitative recovery, for one can never be sure that the whole sample has been dissolved or transferred.

The author has devised an automatic spotter[4] which can apply up to 2 ml volumes of sample or standard and still keep the size of the spot at origin to around 1 cm or less in diameter. Figure 7.2 is a picture of the commercial version of this spotter. Other automatic spotting devices are available, but none with a capacity of 2 ml.

Fig. 7.2. An automatic applicator devised by the Author for thin layer plates.

7.4 DEVELOPMENT

Once the samples and standards have been applied to the thin layer chromatogram, they must be resolved by developing the chromatogram in a suitable solvent system, by a suitable technique. There are four basic techniques for developing chromatograms.

7.4.1 Unsaturated development

This technique involves placing the chromatogram directly into the developing solvent contained in a tank and then covering the tank with a tight-fitting lid.

If multiple solvent systems that contain polar and nonpolar components are used, certain phenomena occur that can enhance or diminish the resolution. As the solvent mixture rises up the plate solvent separation will take place. The more polar components of the developer will be preferentially adsorbed and cause some deactivation of the adsorbent. In addition the polar solvent vapors will condense in the capillaries of the adsorbent so that some liquid–liquid as well as liquid–solid partitioning will take place.

These phenomena reduce the tendency of compounds to form tear-drop shaped spots and tend to produce elliptical spots parallel to the solvent front. This improves resolution and also gives higher R_F values for the more polar substances.

When using unsaturated development, the dimensions of the tank and the development times are restricted. Experience has shown that the depth of the tank should not be much greater than 25 cm and the width 5–9 cm. The development time should be less than it takes for the solvent vapors to reach equilibrium. A 10 cm distance for solvent migration has proved to be optimum. Longer developing times impaired resolution because the plate eventually became deactivated and the migrating solvent started to push the separated components back together. Figure 7.3 shows tank designs and sizes that have proved to be satisfactory.

Fig. 7.3. Tank designs used in the unsaturated development of thin layer chromatograms.

7.4.2 Saturated development

The tank in this case is completely lined with adsorbent paper soaked in the developing solvent mixture. Additional solvent is added to the bottom of the tank and the thin layer plate positioned for development. The tank is then covered with a tight-fitting lid (see Fig. 7.4).

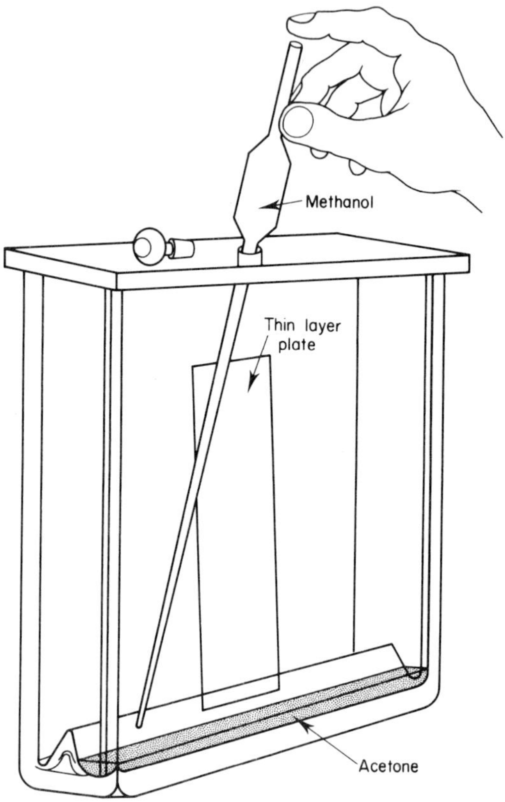

Fig. 7.4. Thin layer chromatography system for the resolution of polar compounds.

The adsorbent paper should be free from any materials which may leach out into the solvent. Thick chromatography paper provides a good material for the lining. Since the tank atmosphere is saturated with solvent vapors the condensation of vapor onto the plate and consequent separation of solvent components is eliminated. The development time is therefore shorter than in an unsaturated system and the spots are more elongated.

7.4.3 Equilibration development

The solvent is placed in a trough on the bottom of the tank and the plate is placed opposite the trough with the adsorbent surface facing it (see Fig. 7.5). The tank is covered with a lid with a hole that can be sealed. The thin layer plate is allowed to equilibrate with solvent vapors for a predetermined time. A solvent of different polarity is then pipetted into the tank through the hole in the lid and development is allowed to proceed. Equilibration allows selective deactivation of the adsorbent and permits the use of a developing solvent of greater or less polarity. Chang[5] used this approach to achieve separation of the

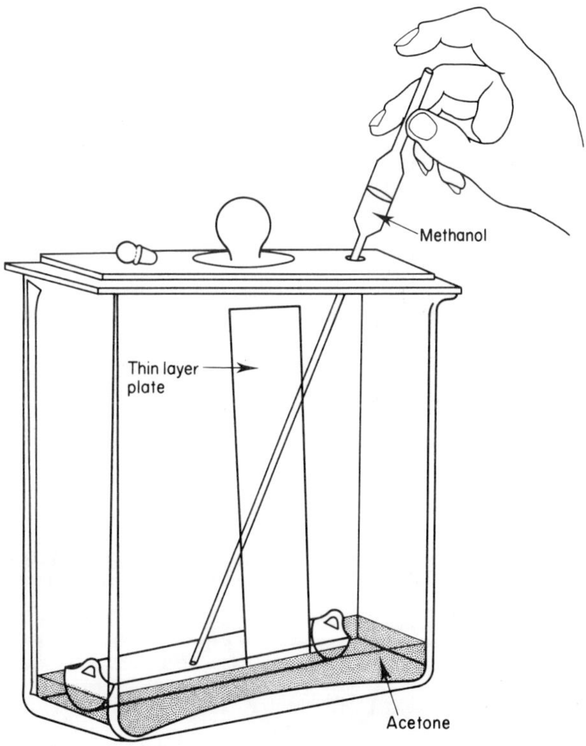

Fig. 7.5. Thin layer chromatography system for the resolution of polar compounds.

glycosidic metabolites of a chemosterilant Dimilan, formed in houseflies. Figures 7.6 and 7.7 show how this technique was used.

This approach was used by the author instead of adding a base or an acid to the developing solvent because of the possibility of salt formation which can produce several spots on a chromatogram even though only one compound is present.

Figure 7.8 shows a commercially available tank designed for the equilibrium technique. The plate is held suspended over the solvent while equilibration takes place and can then be lowered into the developing solvent by means of a teflon screw arrangement.

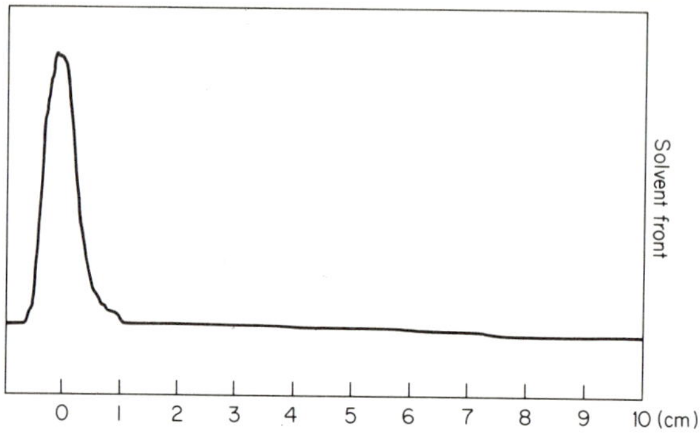

Fig. 7.6. A radiochromatogram of a purified excreta extract from house flies injected with ^{14}C-diflubenzuron (Dimilan). The extract was spotted onto a precoated silica gel sheet (Eastman 13181), and the chromatogram developed with a benzene–dioxane–acetic acid mixture (90:30:1). One compound was detected at origin.

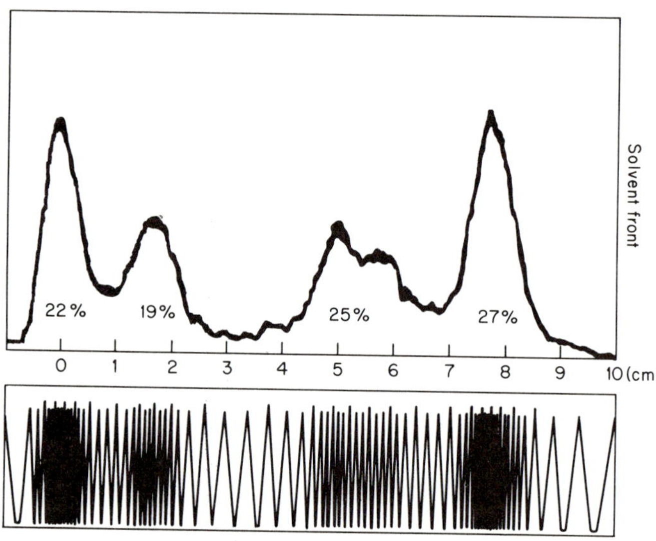

Fig. 7.7. The same material as in Fig. 7.6 was spotted onto a silica gel sheet and deactivated with acetone vapors for 2 h as shown in Figs. 7.4 and 7.5. Methanol was then added through a hole in the tank cover and the plate thus developed. The radiochromatogram showed more compounds to be present in the sample than the one observed in Fig. 7.6.

Fig. 7.8. Commercial tank designed for the equilibrium technique.

7.4.4 Sandwich development

This technique involves placing a spacer between the chromatographic plate and a clean cover plate so that a self-contained tank with the minimum volume is formed.

Figure 7.9 shows a sandwich tank developed by Camag, which uses a plate that has had about 2 mm of adsorbent scraped off from three sides. Cardboard spacers a few mm thick are used to separate the plate from the cover. They are then clamped tightly together and developed in a trough containing the developing solvent.

Fig. 7.9. Camag sandwich development tank.

Figure 7.10 shows a sandwich tank developed by Kodak for their plastic backed precoated sheets. The two glass plates are corrugated to prevent an excessively tight fit but allowing a fit tight enough to hold the sheet in place. The sheet is laid on the first plate, and the second plate is placed over it, exactly flush with the back one. They are then tightly clamped together and developed in a trough containing the solvent.

Fig. 7.10. Kodak sandwich development tank for plastic precoated sheets.

Different analysts may use variations of these various techniques and there is still a great deal of controversy as to which approach is the best. Papers by Geiss *et al.*[6] discuss the merits of saturated development and publications by de Zeeuw[7] discuss the merits of unsaturated development.

In addition to these variations in the type of solvent and the type of tank treatment used for development, the basic chromatographic techniques may be varied by using multiple development, programed multiple development and two-dimensional development, as described in Sections 7.4.5 and 7.4.6.

7.4.5 Multiple development

This technique consists of developing the same plate two or more times in the same direction. The plate is air-dried after each development. This approach has been used to make a compound give a higher R_F value or to move a compound from the origin.

The approach was made more sophisticated by Perry *et al.*[8] who developed a microprocessor-controlled apparatus that could automatically recycle the thin layer plate through a number of preset developments. In each successive development, the solvent was allowed to migrate an increment further. After each development, controlled evaporation allowed the solvent front to recede to any desired position.

This process concentrates the sample in a smaller area and also increases the tendency to form bands, which helps resolution. If the proper parameters are chosen both sensitivity and resolution can be enhanced.

7.4.6 Two dimensional chromatography

This technique is accomplished by developing the plate in one direction, air-drying it and then rotating it through 90° and developing it again with the same or a different solvent system.

A special template described by Mitchell[9] and illustrated in Figs 7.11 and 7.12, develops three standards and one unknown in one direction. The plate is

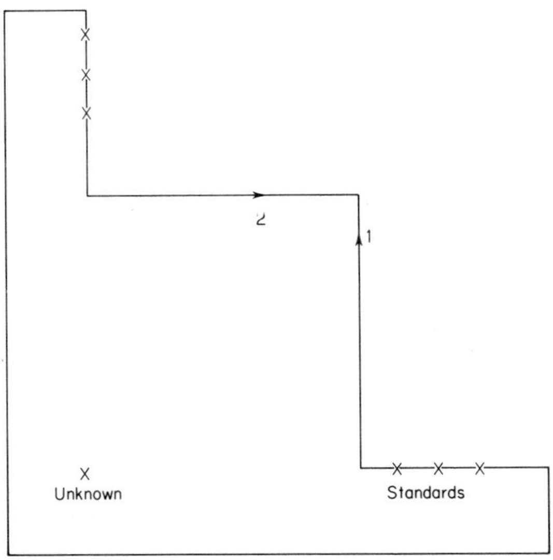

Fig. 7.11. Mitchell's template for two-dimensional chromatography.

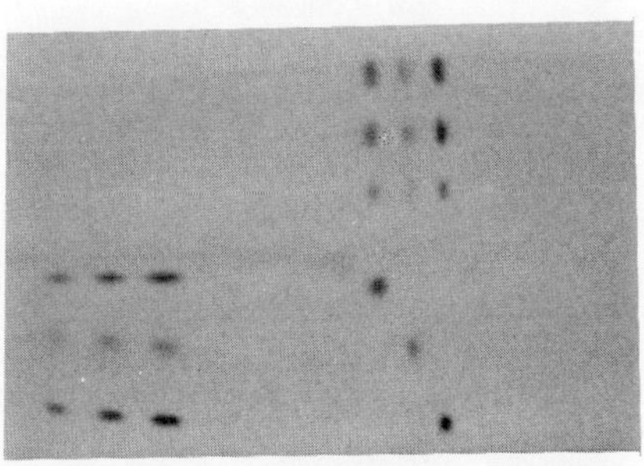

Fig. 7.12. Two dimensional chromatography of a dye mixture dividing the thin layer plate with the template developed by Mitchell (see Ref. 9).

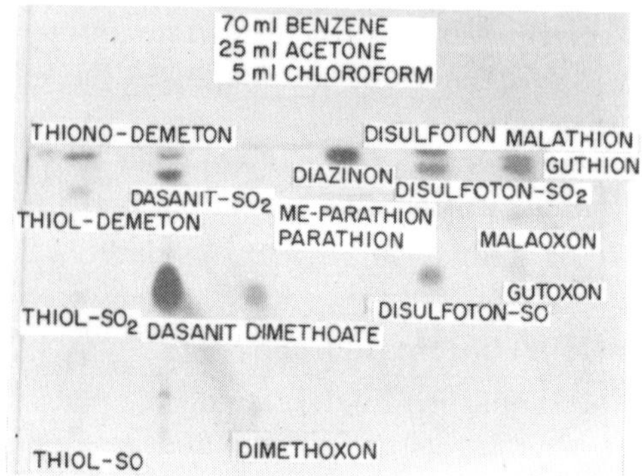

Fig. 7.13. Unsaturated development of some organothiophosphate pesticides with a ternary solvent mixture of 2,2,4-trimethylpentane–acetone–chloroform (70:25:5). Position 1: thiono–demeton, thiol–demeton, thiol–demeton sulfoxide, thiol–demeton sulfone. Position 2: dasanit and dasanit sulfone. Position 3: dimethoate and dimethoxon. Position 4: diazinon, parathion and methyl parathion. Position 5: disulfoton, disulfoton sulfoxide and disulfoton sulfone. Position 6: malathion, malaoxon, guthion and guthoxon.

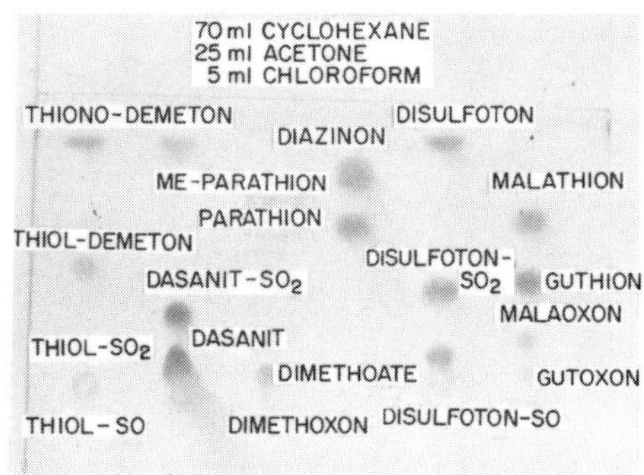

Fig. 7.14. The same series of pesticides chromatographed as in Fig. 7.13, but with hexane replacing the 2,2,4-trimethylpentane.

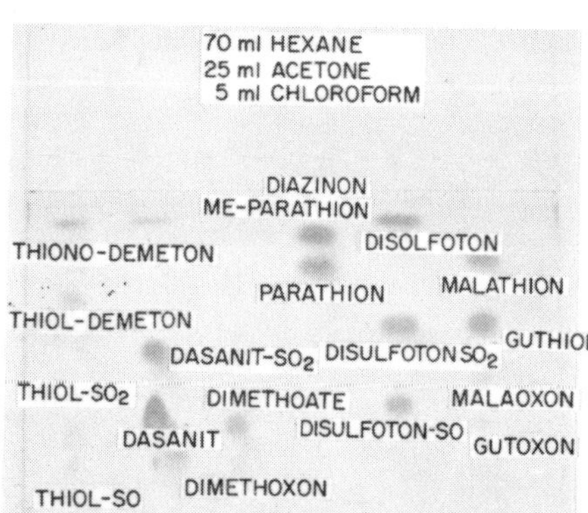

Fig. 7.15. Cyclohexane replaces the saturated hydrocarbon.

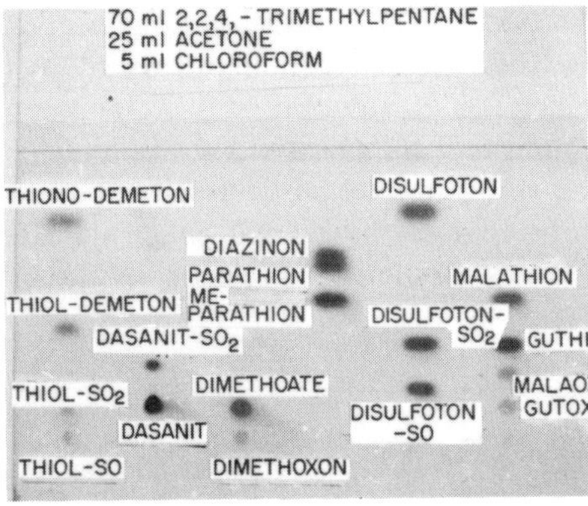

Fig. 7.16. Benzene has been substituted as the hydrocarbon which push the nonpolar pesticides together but improve the resolution of the polar pesticides. This series of plates show how the molecular volume of solvents affect R_F values.

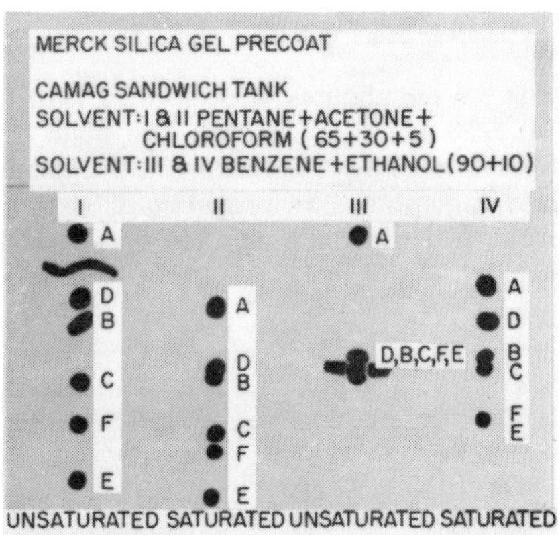

Fig. 7.17. A comparison of the resolution obtained by different solvents and techniques for the pesticide phorate and its five major oxidative metabolites.* A = phorate; B = phoratoxon; C = phorate sulfoxide; D = phorate sulfone; E = phoratoxon sulfoxide; F = phoratoxon sulfone. Position 1 represents the resolution obtained with pentane–acetone–chloroform (65:30:5) used in an unsaturated system. Position 2 represents the resolution under saturated conditions. Position 3 shows the resolution obtained in a sandwich tank using a solvent mixture of benzene–ethanol (90:10). Position 4 shows the R_F positions obtained when the cover plate was replaced with a cellulose thin layer plate. The cellulose was saturated with the solvent and then placed facing the silica gel layer.*

* See footnote p. 139.

removed, air-dried and then three more standards are spotted. The plate is developed in a direction at 90° to the first development. This enhances identification because the two sets of standards make the identification of any compounds with coinciding R_F values doubly certain.

The author has found that the unsaturated technique gave excellent results when applied to the identification of pesticide residues. Figures 7.13–7.15 show how R_F values may be changed by changing the saturated hydrocarbon, and Figs 7.17–7.20* show how resolution is affected by the different development techniques.

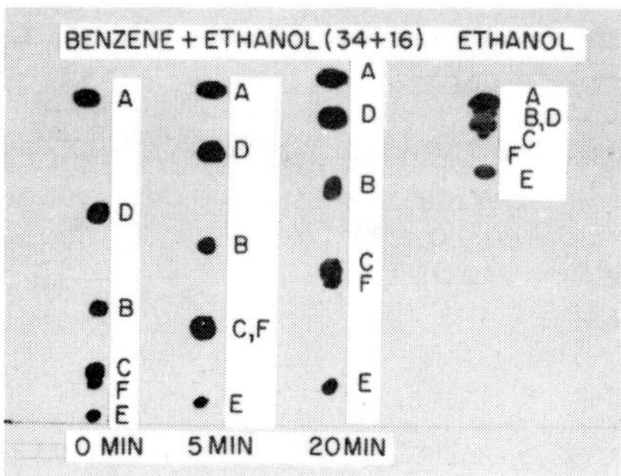

Fig. 7.18. Phorate and its metabolites were spotted onto the thin layer plate and equilibrated for 0, 5 and 20 min with ethanol. The plate was then lowered into the developing solvent of benzene–ethanol (34:16). Another plate was developed just with ethanol. These chromatograms show how proper equilibration with polar solvent vapor before development can help to control resolution.*

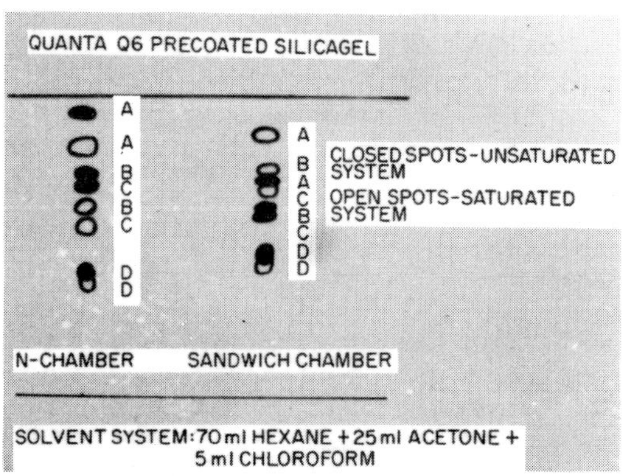

Fig. 7.19. Phorate and its metabolites were chromatographed with a ternary solvent system of hexane–acetone–chloroform (70:25:5) in an N-chamber, and in a sandwich chamber with and without saturation.*

* See footnote p. 139.

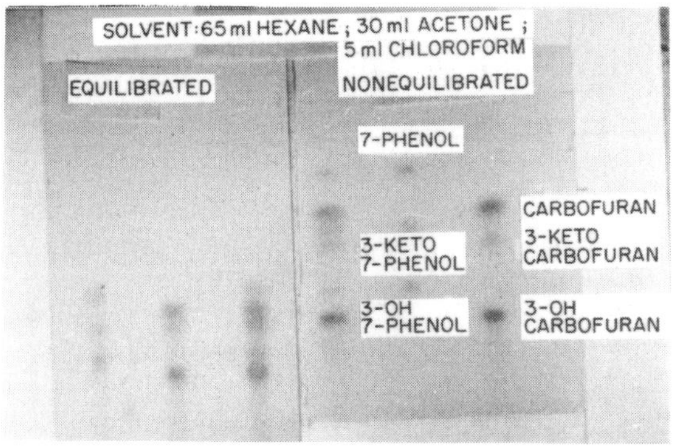

Fig. 7.20. Carbofuran and some of its metabolites. 3-Keto carbofuran, 3-hydroxy carbofuran, 7-phenol analog, 3-keto-7-phenol analog, and the 3-hydroxy-7-phenol analog were spotted on one side of a thin layer plate equilibrated with a ternary solvent mixture of hexane–acetone–chloroform (65:30:5). After 2 h equilibration it was developed in the same solvent. After development, enough of the adsorbent was scraped off so that solvent would not be adsorbed on that side. The other side was developed in a fresh mixture of the same solvents without equilibration.*

7.5 DETECTION SYSTEMS

Spots which have migrated on a thin layer chromatogram have to be made visible in order to calculate R_F values and to make comparisons with standards. This is done by reacting them with chemical solutions to produce colored areas or by observation under filtered UV illumination to reveal naturally fluorescent compounds. A fluorescent material may also be incorporated into the thin layer so that the migrated compounds can be detected as fluorescence-quenching dark spots against a bright fluorescent background.

Detection reagents are usually applied as solutions which can be sprayed onto the plate, or the plate is dipped into special tanks containing the reagent solutions. Figure 7.21 shows an air pressure spray bottle and an aerosol spray combination used for applying reagents. Figure 7.22 shows two types of tanks for dipping the plates. One is a narrow glass sandwich type tank and the other is a stainless steel tank that was originally used by Mitchell[9] for impregnating immobile phases into paper chromatograms.

General charring reagents can be compared to the flame ionization detector used in gas chromatography. The response is dependent upon the carbon content of the molecule. The following solutions can be applied as charring reagents.[10]

* Figures 7.17–7.20 are tracings of chromatograms from the unpublished work of M. E. Getz, Analytical Chemistry Laboratory, Agricultural Environmental Quality Institute, SEA, US Department of Agriculture, Beltsville, Md. USA, and J. Sherma, Chemistry Dept, Lafayette College, Easton, Pa. USA.

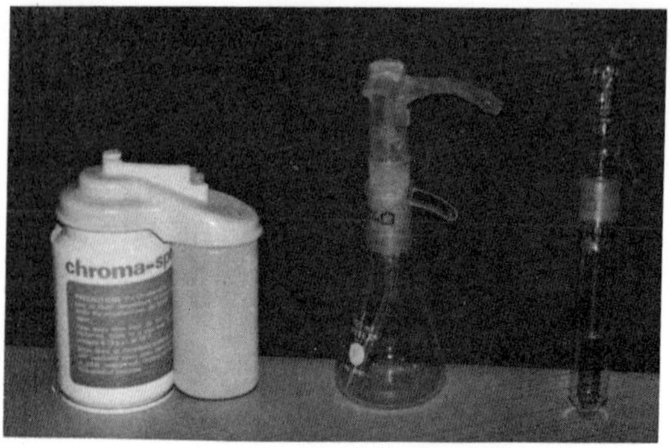

Fig. 7.21. Typical reagent spray applicators.

Fig. 7.22. Tanks designed for the application of detection reagents.

(i) Spray the plate with or dip it into a 20% solution of sulfuric acid in ethanol and heat to 100–110 °C for a few minutes. Organic compounds will appear as black spots.

(ii) Dissolve 3 g of sodium dichromate in 20 ml of water and add 10 ml of concentrated sulfuric acid. Heat to 100–110 °C. Black spots will appear.

(iii) Spray with a saturated solution of potassium dichromate in concentrated sulfuric acid and heat to 110 °C for a few minutes. Black spots develop.

(iv) Spray with a solution of 5% nitric acid in concentrated sulfuric acid. Heat to 110 °C for a few minutes. Black spots are formed.

Some other general detection reagents are:

(i) A 3% solution of cupric acetate dissolved in an 8% solution of phosphoric acid, sprayed onto the plate or the plate dipped into it. The plate is heated at 130 °C for 30 min. Compounds which have at least one double bond or hydroxyl group react to give a rusty brown spot.

(ii) A 1% solution of iodine in chloroform or ethanol will react with many compounds to give brown spots, although in many cases this is a reversible reaction and the spots are not stable. Another approach is to place the plate in a chamber containing iodine vapors.

(iii) The plates can be treated with a 1–2% solution of *N*-bromosuccinimide, or placed in an atmosphere of bromine vapors. They are then treated with an ethanolic solution of fluorescein. Fluorescein reacts with the bromine to form eosin which fluoresces greenish-yellow under UV illumination; this is the background. Where compounds are present they will absorb bromine and no eosin will form so that they are detected as dark quenched areas against the fluorescent background.

(iv) Reactions with sulfuric and nitric acid can produce some very selective colorations. They must be determined by experimentation since they cannot be predicted. The chromatogram is dipped into or sprayed with a solution of the acid in water or alcohol. The plate is placed in a forced-draft oven or on a hot plate under a hood. The initial temperature of the oven or hot plate is 50 °C and is raised in 10° increments; the plate is allowed to remain at each temperature for several minutes. If no colors are visible, the plate is illuminated with a hand-held filtered UV lamp to check for induced fluorescence. Once the temperature gets over 100 °C, charring will take place.

7.6 COUPLING COLOR REACTIONS

A simple and efficient way to produce selective colors is to utilize the property of molecular coupling or condensation. In order to obtain a colored compound, the new molecule formed must contain a chromophoric group. Important chromophoric groups are: —N═N— (azo); $—NO_2$ (nitro); —N═O (nitroso); —C═O (carbonyl); C≡N (cyano); $═C_6H_4═$ (*p*-quinoid); —C═C—C═O (unsaturated ketone); $—CH_2—$ (methine); —N═C— (azo methine); —N≡N (diazonium); —NH—CO (carboxylamide); $—SO_2—N═$ (sulfonamide); $—CO—CH_2—CO—$ (acetoyl-acetyl).

A well-known coupling reaction is that of the diazonium salts with an aromatic molecule containing an amine, substituted amine or phenol group. If a compound has a primary amine configuration on an aromatic molecule, it can be diazotized and will become capable of coupling with other compounds to form deeply colored molecules. Diazotization proceeds as follows:

$$Ar—NH_2 + HNO_2 + HX \rightarrow Ar—N_2{}^+X^- + 2HOH$$

The coupling reaction is:

$$Ar{-}N_2^+ + Ar'{-}H \rightarrow Ar{-}N{=}N{-}Ar' + H^+$$

Because there is a displaced hydrogen in the coupling reaction it takes place more efficiently under neutral or slightly alkaline conditions.

A chromatographed compound containing a diazotizable amino group can be converted to a diazonium salt on a thin layer plate by treating with a solution of sodium nitrite and concentrated hydrochloric acid. Excess nitrous acid is destroyed by treating the plate with urea solution. The plate can then be treated with a suitable coupling agent to form a colored compound.

Two widely used diazonium salts are *p*-nitrobenzenediazonium fluoroborate and diazotized sulfanilic acid. Two other salts worth investigating are 1-(*p*-*n*-decyloxyphenyldiazonium)chloride which is less soluble in water than in organic solvents,[11] and bis(2-diazonium-5-nitrophenyl)ether which can couple at pH 5.[12] Another widely used coupling reagent is *N*-(1-naphthyl)ethylenediamine, commonly known as Bratton-Marshall reagent.[13]

Sawicki *et al.*[14] found that polycyclic compounds like azulene could couple with diazonium salts. Diazonium salts can also couple with active methylene groups such as those contained in methyl-1,3-indandione-2-carboxylate.[15] Compounds having displaceable hydrogen in the *ortho* or *para* positions may also be capable of coupling with diazonium salts.

Oxidative coupling reactions can also produce a range of colored compounds and they form the basis of the modern color photography processes. Indeed, chromatography draws on the chemistry of color photography for many suitable reagents.[16] Usually atmospheric oxygen is suitable for initiating the reaction, but sometimes the kinetics are too slow and then additional treatment of the plate, with a dilute solution of peroxide or an oxidizing salt is necessary. Some background color may form but the spots are generally much more intense than the background. Some typical reactions are:

Reactions of *p*-aminophenol with phenols to produce indophenol dyes.

$$H_2N{-}C_6H_4{-}OH + Ar{-}OH \rightarrow HO{-}C_6H_4{-}N{=}Ar{=}O + HOH$$

p-Phenylenediamine reacts with phenols to form an indoaniline dye.

$$H_2N{-}C_6H_4{-}NH_2 + Ar{-}OH \rightarrow H_2N{-}C_6H_4{-}N{=}Ar{=}O + 3H^+$$

Amines couple with *p*-phenylenediamine to form indamine dyes.

$$H_2N{-}C_6H_4{-}NH_2 + Ar{-}NH_2 \rightarrow H_2N{-}C_6H_4{-}N{=}Ar{=}NH + 3H^+$$

The reaction of *p*-phenylenediamine with thiopenols produces indothiophenol dyes.

$$H_2N{-}C_6H_4{-}NH_2 + Ar{-}SH \rightarrow HS{-}Ar{-}N{=}C_6H_4{=}NH + 2H^+$$

Compounds containing active methylene groups will couple with *p*-aminophenols or *p*-phenylenediamines to form azomethine dyes.

$$H_2N—C_6H_4—NH_2 + R_1—CH_2—R_2 \rightarrow H_2N—C_6H_4—N{=}C(R_1)(R_2) + 4H^+$$

The methylene group is active if it is found adjacent to polar groups such as —CN, =CO, CH_3CO, $COOC_2H_5$ and C_6H_5.

Other amines useful for coupling are *p*-toluylenediamine and *p*-diethylaminoaniline. The addition of a halogen to the *p*-aminophenol molecule increases its coupling capacity and the diamines may be substituted, but one free amino group must be present for coupling purposes. Coupling usually takes place by displacing a hydrogen atom, in the *ortho* or *para* position, to the OH or NH_2 group of the coupling agent. This holds true for coupling with both diazonium compounds and quinoneimides.

Certain heterocyclic compounds such as 1-phenyl-3-methyl-4-amino-5-pyrazolone are able to couple as effectively as compounds having a nitrogen atom in the ring if there is a CH_3 group in the α position to the N atom. Other useful coupling reactions are those of 9-chloroacridine with the primary aromatic amines;[17] 3-methyl-2-benzothiazoline with the phenols;[18] 2,6-dichloroquinone-4-chloroimide for the phenols, substituted anilines and naphthylamines, aromatic nitrogen heterocyclics and some polynuclear hydrocarbons.[19] Alkylating reagents such as epoxides, alkyl halides and organophosphate pesticides can be detected by their reactions with *p*-nitrobenzyl pyridine.[20]

It may appear unnecessary to have such a wide choice of reagents at one's disposal for reaction with essentially similar molecules, but the positive identification of a compound in an environmental sample for legal purposes requires as much data as possible to be conclusive. Some substituted molecules may only react with certain types of coupling agents because of the nature of the adsorbent, a situation which can be overcome by the availability of a wide range of reagents.

Most lists of available chromogenic reagents do not explain how they work, so if a reagent produces no positive reaction, it is extremely difficult to work out why without any knowledge of the reaction mechanism. A reagent may be known to produce orange colors, and may be mistakenly taken as showing a positive reaction because an interference is present that also gives orange colors. Switching to another coupling agent may give blue colors with the compound to be assayed and a completely different color with the interference, thereby providing definite proof of the presence of the compound.

7.7 PYROLYSIS AND HIGH TEMPERATURE REACTIONS

As mentioned previously, aluminum-backed precoated thin layer plates are commercially available. This opens up the technique of pyrolysis and high temperature reactions for the detection of compounds on thin layer plates, because aluminum plates can tolerate temperatures in excess of 200 °C.

Feigl's book on spot tests[21] contains a wealth of reactions that could be applied to chromatographic detection. Many of these require pyrolysis and heat to produce color reactions. The reactions and reagents described are quite similar to those enumerated in the appendixes of many chromatography books, but here, in addition, the mechanism of the reaction is described where it is known.

The author presents here the reactions that he believes can be adapted for chromogenic identification on thin layer plates. Some have already been mentioned but others have not. Since these tests were originally devised for use with small volumes of solvents in micro test tubes, or in small reservoirs in porcelain plates, or for detecting vapors, some modifications are necessary for adaptation to thin layer chromatography. Several general rules should be observed.

(1) If the reagent is to be applied as a spray the concentration should usually be between 2 and 10 %.
(2) If the reagent is to be applied by dipping the plate into the solution, the concentration may be lowered to 0.2–2 %.
(3) Dipping reagents should be dissolved, if possible, in solvents which do not dissolve the chromatographed compounds.
(4) Gloves should be used, and all treatments performed in a hood if possible, or in a well-ventilated area.

7.7.1 Pyrolysis reactions

Dry heating of organic molecules forms end products which indicate that a hydrolytic process has taken place. The water necessary for hydrolysis may originate from the oxygen and hydrogen in the compound itself or from a donating compound which can be impregnated into the thin layer adsorbent.

A simple way of producing these pyrolytic reactions on thin layer plates is to place a thick aluminum plate on top of a hot plate, with a suitable hole for inserting a thermometer with which to monitor the temperature. The chromatogram may then be placed on the plate to be heated.

Useful water-donating compounds are: oxalic acid dihydrate which releases water at 80–150°; succinic and phthalic acids at 220 °C; manganous sulfate at 100–200 °C. Succinic and phthalic acids will form anhydrides on losing water, and are then capable of reacting with certain amines and carbamates to form fluorescent compounds.[22] Aniline can be formed from anilides, phenol from phenol esters, and hydrogen sulfide from thioketones if hydrolysis takes place in the presence of a source of ammonia such as urea, biuret or guanidine carbonate. The reactions are:

$$ArCONHC_6H_5 + NH_3 \xrightarrow{\Delta} ArCONH_2 + C_6H_5NH_2$$
$$ArCOOC_6H_5 + NH_3 \xrightarrow{\Delta} ArCONH_2 + C_6H_5OH$$

The aniline or phenol formed can be coupled with a diazonium salt or oxidatively coupled to form a colored compound.

$$(Ar)(R)—C{=}S + NH_3 \xrightarrow{\Delta} (Ar)(R)—C{=}NH + H_2S$$

The hydrogen sulfide formed may be reacted with lead or silver if the appropriate salt has been incorporated into the thin layer either before or after development. The lead or silver sulfide distinguishes the migrated compounds as brown or black spots.

7.7.2 High temperature reactions

If a thin layer plate is treated with a solution of benzoin and then heated to 120–160 °C, the benzoin releases active hydrogen and converts to benzil. The following reactions could take place.[23]

$$RX + 2H \rightarrow RH + HX\,(X = \text{halogen}, SO_3H, NO_2, SH; R = \text{aliphatic group})$$

$$RN—NO_2 + 2H \rightarrow RNH + HNO_2$$

$$RONO_2 + 2H \rightarrow ROH + HNO_2$$

$$RCNS + 2H \rightarrow RCN + H_2S$$

Benzoyl peroxide, when fused at 110–130 °C releases oxygen and can remove *O*-ethyl and *N*-ethyl groups.

$$R—OC_2H_5 + O \rightarrow ROH + CH_3CHO$$

$$R—NHC_2H_5 + \rightarrow RNH_2 + CH_3CHO$$

O-methyl and *N*-methyl groups behave in a similar manner, except that formaldehyde is released instead of acetaldehyde.

Sulfur-containing compounds may be detected by heating the benzoin-treated plate to 150 °C. Hydrogen sulfide is released, and if a lead or silver salt is also impregnated into the adsorbent black or brown spots will appear.[24]

Many nitrogen-containing organic compounds release nitrous oxide when they are heated with an excess of manganese dioxide[25] which can be introduced by incorporating soluble manganese salts which decompose at elevated temperatures to form the dioxide. The nitrous oxide will react as nitrous acid and diazotize a compound such as sulfanilic acid which could then be coupled to form an azo dye.

Organophosphate pesticides can be converted to the orthophosphate ion by heating with ammonium persulfate.[26] The orthophosphate ion is reacted with molybdate, which is then reduced to molybdenum blue and this distinguishes the pesticides as blue spots against a white or yellow background.

Alkyl, aryl and acyl amines react with fluorescein chloride and zinc chloride at elevated temperatures to produce fluorescent rhodamine dyes.[27] The plate should be treated with a solution of fluorescein chloride and zinc chloride and then heated to 250–260 °C. Viewing the plate under filtered UV illumination reveals the compounds as fluorescent spots at sites of positive reaction.

Tertiary amines and quaternary amines will react with citric or *cis*-aconitic acid in the presence of acetic anhydride to form red to violet colors.[28] A 2% solution of citric or *cis*-aconitic acid in acetic anhydride is made, and the plate is dipped into this solution and heated at 80 °C.

N-Arylamides can be nitrosated with nitrous acid, converted to diazotates by reaction with alkali and then coupled with α-naphthol to give red-violet compounds.[29] Other phenols may also be used for coupling.

$$\mathrm{Ar{-}NH{-}COR + HNO_2 \rightarrow Ar{-}N(NO){-}COR + HOH}$$

$$\mathrm{Ar{-}N(NO){-}COR + 2KOH \rightarrow Ar{-}N{=}N{-}OK + RCOOK + HOH}$$

$$\mathrm{Ar{-}N{=}N{-}OK + Ar{-}OH \rightarrow Ar{-}N{=}N{-}Ar{-}OH + KOH}$$

Nitrosation may be accomplished by treating the plate with a 10% sodium nitrite solution in concentrated hydrochloric acid, then with 0.1% urea to destroy the excess nitrous acid, and finally with 10N alkali to form the diazotate.

Nitriles can be heated to 180 °C with hydrated oxalic acid to produce amides, or reacted again with oxalic acid to form oxamide which can then be heated with thiobarbituric acid to give an orange compound.[30] The treatment with oxalic acid requires a temperature of 200 °C which is higher than that required by the reaction with the hydrate. The treatment with thiobarbituric acid requires heating at 140–160 °C.

Hydrazine derivatives of organic compounds can react with glutaconic aldehyde to form red to orange compounds.[31] The glutaconic aldehyde can be made *in situ* on the plate by first treating it with 4-pyridylpyridinium dichloride and then alkali.[32] A 1% solution of pyridylpyridinium chloride plus 1N sodium hydroxide is used.

7.8 OTHER CHROMOGENIC REACTIONS

The many other chromogenic reactions that could be applied to thin layer chromatograms are as follows:

Aromatic aldehydes and α and β unsaturated aldehydes react with thiobarbituric and phosphoric acids at 120–140 °C to give orange condensation products.[33]

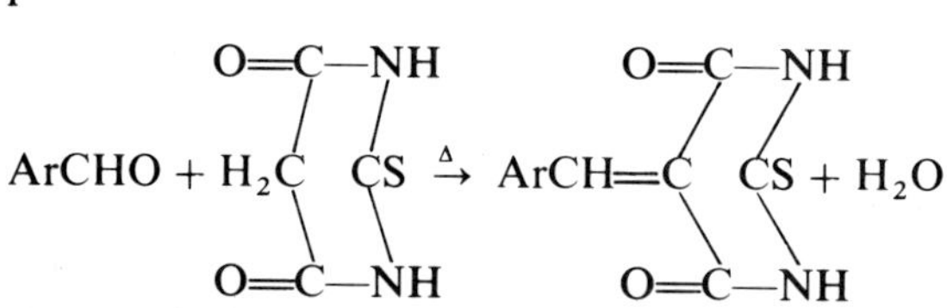

Aromatic aldehydes also condense with *N*,*N*-dimethyl-*p*-phenylenediamine in the presence of hydrochloric or oxalic acid to form Schiff bases which protonate to give red colors.[34]

$$ArCHO + NH_2C_6H_4N(CH_3)_2 \rightarrow ArCH{=}NC_6H_4N(CH_3)_2 + HOH$$

$$ArCH{=}NC_6H_4N(CH_3)_2 + H^+ \rightarrow ArCH{=}NHC_6H_4N(CH_3)_2{}^+$$

Phloroglucinol condenses with aromatic aldehydes in the presence of hydrochloric acid to give orange-pink colors.[35]

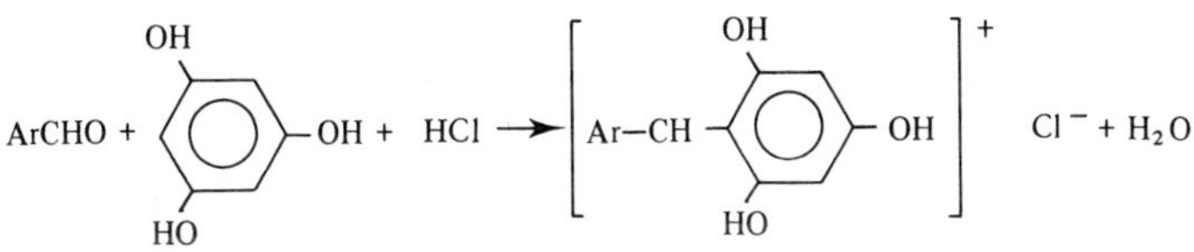

Acidic solutions of azobenzenephenylhydrazinesulfonic acid react with aldehydes or ketones to give red or blue-colored compounds.[36]

$$C_6H_5{-}N{=}N{-}C_6H_4{-}NH{-}NHSO_3H + (R)(Ar)CHO$$
$$\downarrow$$
$$C_6H_5{-}NH{-}N{=}C_6H_4{=}N{-}N{-}CH(R)(Ar) + H_2SO_4$$

An organic solvent solution of the reagent containing hydrochloric or sulfuric acid should be used and the plate heated to 100–110 °C.

Aromatic and α,β-unsaturated carbonyl compounds condense with *p*-nitrophenylhydrazine to yield Schiff bases which can then be reacted with alkali to form blue or violet colored compounds.[37]

$$-C{=}C{-}C{=}O + H_2N{-}NH{-}C_6H_4{-}NO_2 \rightarrow -C{=}C{-}C{=}N{-}NH{-}C_6H_4{-}NO_2 + H_2O$$

The plate can be treated with a solution of *p*-nitrophenylhydrazine in alcohol and hydrochloric acid and then heated to 110 °C. Subsequent treatment with an alcoholic base solution develops the color.

Primary aromatic amines condense with aliphatic and aromatic aldehydes to form highly colored Schiff bases.[38] The chromatogram is treated with an alcoholic solution of *o*-dianisidine or 2,7-diaminofluorene. The plate is then heated to produce the reaction. The colors formed depend upon the structure of the aldehyde.

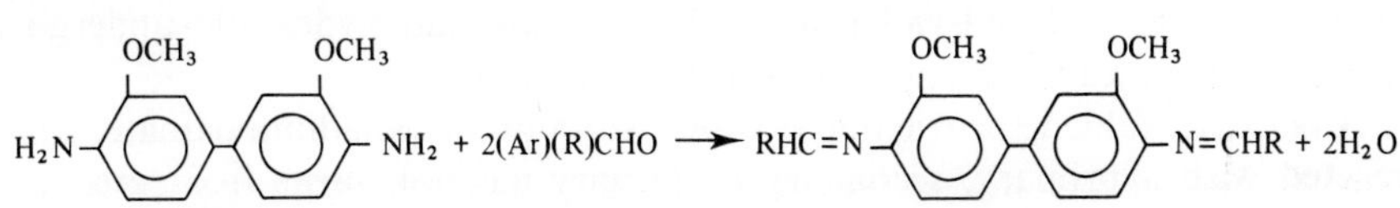

Organic compounds which exhibit oxidizing properties can be detected by reacting with thio-Michler's ketone (4,4′-bis-dimethylaminothio benzophenone) or tetrabase (tetramethyldiaminodiphenylmethane). Blue quinoidal structures are formed when these compounds are oxidized.[39]

$$(CH_3)_2N{-}C_6H_4{-}CH_2{-}C_6H_4{-}N(CH_3)_2 + O \rightarrow (CH_3)_2N{-}C_6H_4{-}CH{=}C_6H_4{=}N^+(CH_3)_2 + OH^-$$

Halogenated compounds, the organochlorine pesticides in particular, can be detected by their reaction with aromatic amines after exposure to high intensity UV radiation. Adamovic[40] published a paper listing 18 aromatic amines which will react with organochlorine compounds under UV illumination to produce colored spots. The likely mechanism of reaction is that the UV radiation strips the halogen from the molecule and this then oxidizes the amine to a Wurster-type dye.

$$H_2N{-}C_6H_4{-}N(CH_3)_2 + 2Cl \rightarrow NH{=}C_6H_4{=}N^+(CH_3)_2 + 2Cl^- + H^+$$

The reaction resembles that obtained with the oxidizing compounds except that UV radiation is the initiator of the reaction.

Le Rosen *et al.*[41] found that concentrated sulfuric acid containing a low concentration of formaldehyde (0.2%) would react with aromatic hydrocarbons, phenols, polyphenols, thiophene and many other compounds at room temperature or with gentle heating to form a variety of colors.

Some phenolic ethers and esters will form phenols when strongly heated.[42] The resulting phenols may then be detected by the various reagents available. The correct temperature for 'pyro-hydrolysis' of the ether or ester should be determined experimentally.

Ehrlich[43] discovered in 1882 that diazotized sulfanilic acid would couple with imidazole derivatives at the *α* position to the imide group. This proved to be a useful test for histamine-type compounds.

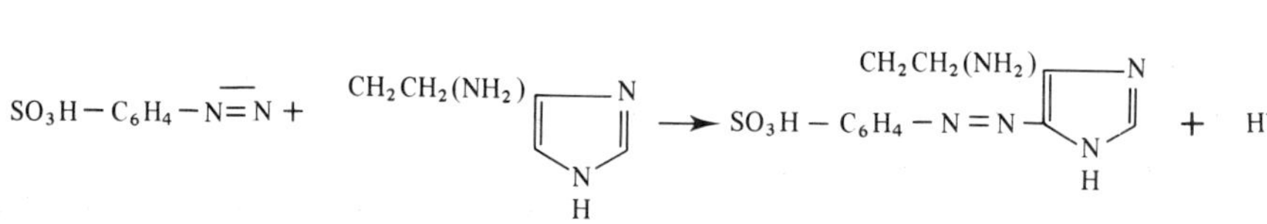

Ehrlich and Herter[44] demonstrated that a neutral or slightly alkaline solution of sodium 1,2-naphthoquinone-4-sulfonate was able to couple with reactive methine and amino groups.

Compounds containing the allyl group react with phloroglucinol in hydrochloric acid to give red-violet colors.[45] Aromatic aldehydes will undergo a similar condensation reaction to form trimethine dyes.

A benzene, toluene or trichloroethylene solution of vanadium oxinate, when reacted with a primary, secondary or tertiary alcohol, turns from greenish-gray to red.[46] Vanadium oxinate can be made by mixing 1 ml of 2.5% 8-hydroxyquinoline in 6% acetic acid with every milligram of a vanadium salt solution. The mixture is then shaken with the desired solvent.

β-Naphthol can be detected in the presence of α-naphthol by its reaction with 2-nitroso-l-naphthol dissolved in concentrated nitric acid.[47] There is no reaction with the α-isomer, but a red-violet compound is formed with the β-isomer.

A strongly-acid aqueous solution of azobenzenephenylhydrazinesulfonic acid reacts with aldehydes to give red or blue colors.[48] Aromatic aldehydes give red colors and aliphatic aldehydes blue ones. Some ketones may react in a similar manner, but the reaction kinetics are much slower.

Aromatic α,β-unsaturated carbonyl and 1,2-dicarbonyl compounds react with *p*-nitrophenylhydrazine to form Schiff bases and then will react with alkali to form violet or blue salts.[49]

$$C{=}C{-}C{=}O + H_2N{-}NH{-}C_6H_4{-}NO_2 \rightarrow C{=}C{-}C{=}N{-}NH{-}C_6H_4{-}NO_2 + HOH$$

$$C{=}C{-}C{=}N{-}NH{-}C_6H_4{-}NO_2 + KOH \rightarrow C{=}C{-}C{=}N{-}N{=}C_6H_4{=}NO_2K + HOH$$

Sulfurous acid decolorizes *p*-fuchsin[50] and other triphenylmethane dyes.[51] These decolorized leuco bases will react with aromatic or aliphatic aldehydes to form deeply colored products. The decolorization of the dye solution can be accomplished by bubbling sulfur dioxide through it until the color disappears. A 0.1% dye solution would be suitable.

When indole is reacted with aldehydes in hydrochloric acid solution, red compounds are formed.[52] The solutions have to be freshly prepared because they oxidize easily in air.

Carboxylic acids can be detected as ferric hydroxamates.[53] The acid must first be changed to the acid chloride by reaction with thionyl chloride.

$$ROOH + SOCl_2 \rightarrow RCOCl + SO_2 + HCl$$

The acid chloride is then reacted with a basic solution of hydroxylamine to form the hydroxamate.

$$ROCl + NH_2OH + 2NaOH \rightarrow RCO(NHONa) + NaCl + 2HOH$$

The hydroxamate reacts with the ferric ion to form a complex salt.

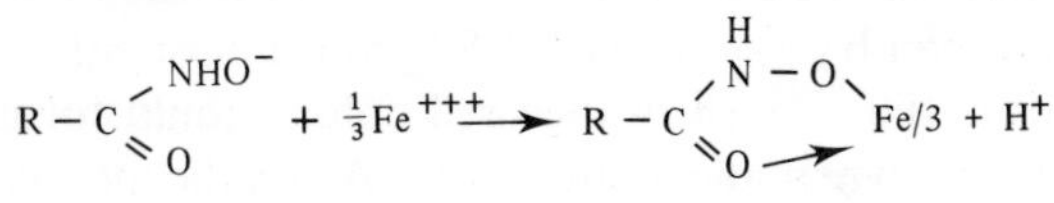

The acid chloride can be made by treating the chromatogram with thionyl chloride solution under a fume-hood. After the reaction is complete, excess chloride can be driven off by gentle heat. The plate is then treated with an alkaline solution of hydroxylamine in alcohol in order to convert it to the hydroxamate. Subsequent reaction with a ferric chloride solution produces reddish-purple spots.

Esters of carboxylic acid can be detected in a similar manner,[53] though they do not have to be reacted with the thionyl chloride. Anhydrides and imides of carboxylic acids will give a positive reaction for this test.

Thiol compounds can be detected by reacting them with cuprous acetate solution or with cupric acetate solution containing hydroxylamine which reduces the cupric form to the cuprous.[54] Cuprous salts of mercaptans form:

$$2Cu^+ + 2RS^- \rightarrow Cu_2(RS)_2$$

Sulfones can be detected by the pyrolytic release of sulfur dioxide.[55] The sulfur dioxide reacts with stannous hydroxide to form the black compound hydrated nickel dioxide [$NiO(OH)_2$]. After chromatography the stannous hydroxide can be produced on the plate by treating it with a solution of nickel sulfate and then with 1N ammonium hydroxide. The stannous salt is then precipitated uniformly throughout the adsorbent. When the sulfur dioxide is released at high temperature, black spots become visible at the sites containing the sulfones.

Nitroso compounds react with phenol and concentrated sulfuric acid to give a red color which turns blue in alkaline conditions.[56] They also react with sodium pentacyanoamminoferroate ($Na_3(Fe(CN)_5NH_3)$) to give green or violet colors.[57] *N,N′*-Diphenylbenzidine in concentrated sulfuric acid solution is converted to blue *p*-quinoidal compounds by nitrates, nitrites and other oxidants, as well as by nitroso compounds.[58]

$$ArNO + C_6H_5{-}NH{-}C_6H_4{-}C_6H_4{-}NH{-}C_6H_5 \rightarrow ArNHOH + C_6H_5N{=}C_6H_4{=}C_6H_4{-}C_6H_5$$

Nitrosamines can be detected by releasing the nitrous acid hydrolytically. They are heated with a solution of sulfanilic acid and 1-naphthylamine in acetic acid. Concentrated hydrochloric acid is sometimes required to be present before the reaction takes place.[59]

Aliphatic nitrosamines can be detected by denitrosating them with hydrazoic acid to form secondary amines which can then be reacted with carbon disulfide and an ammoniacal copper solution to produce brown copper dithiocarbamates.[60]

Nitro compounds can be detected by fusion with tetrabase or with diphenylamine.[61] The plate can be treated with a 5% solution of either base in benzene and then heated at 100 °C. An orange red color should form, though it can sometimes disappear when the plate cools. Aromatic polynitro compounds release nitrite when heated with concentrated alkali.[62] The released nitrite can then be reacted with Griess reagent (sulfanilic acid).

Organic thiocyanates form blue compounds when heated with sodium sulfite and ferric chloride:[63] the plate is treated with a solution of sodium sulfite in sodium hydroxide, heated, and then treated with ferric chloride in sulfuric acid solution.

Phenoxy compounds capable of releasing phenol can be detected by reacting them with hydrazoic acid and *N,N'*-diphenylbenzidine.[64] The plate is treated with sodium azide, then with a freshly prepared 0.05% solution of the benzidine in concentrated sulfuric acid.

All the reagents described can be very useful for developing selective chromogenic reactions. Optimum reaction conditions for the particular problem in hand may only be determined experimentally.

To illustrate the need for knowing the nature of any chemical reaction being used, Dragendorff's reagent and Gibbs' reagent are considered here.[65]

Dragendorff's reagent and many of its modifications have long been used for detecting alkaloids and other compounds containing tertiary or quaternary nitrogen groups. (The reaction was discovered in 1889 by Leger.[66]) The reagent consists of a solution of bismuth subnitrate ($4BiNO_3(OH)_2 . BiO(OH)$), and potassium iodide. The bismuth salt is not soluble in water or organic solvents but it is soluble in hydrochloric, acetic and sulfuric acids. This accounts for different acids being used in the many modifications. When a weakly acid solution of the bismuth salt and potassium iodide is reacted with the active nitrogen-containing compounds, reddish-orange double iodides are formed with the structure: $BiI_3 . NC . HI$ (NC = heterocyclic compound).

Gibb's reagent,[65] 2,6-dibromo- or 2,6-dichloroquinone chlorimide (TCQ), was used for forming colored indophenol compounds with active phenols. However, this reagent can be used in two different ways depending upon the reaction conditions. In the presence of hydrochloric acid the chlorimide reacts with sulfur-containing compounds to form *N*-sulfenyl-2,6-dihaloquinone imines.

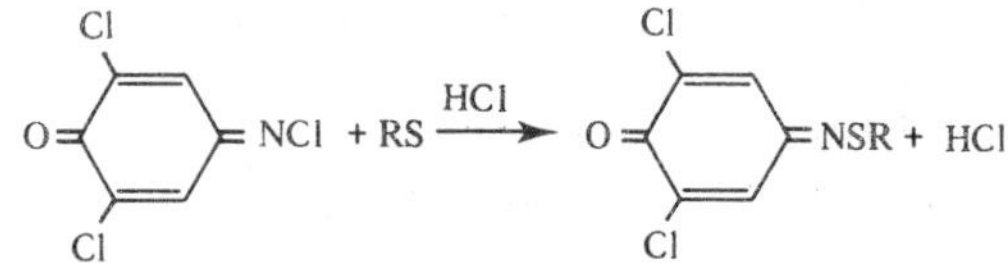

Cuzzoni and Pietra Lissi[67] characterized the reaction and found that color formation was related to structure. When R = CH_2Ar the color is orange, when R = Ar the color is red, when R = alkyl the color is yellow.

This reagent has been used for detecting organothiophosphate pesticides,[68] in the form of a 0.5% solution of the 2,6-dichloro compound in a saturated hydrocarbon such as hexane or 2,2,4-trimethylpentane. The solution is extremely stable under these conditions. If, however, it is dissolved in a polar solvent such as methanol or acetone its active life is only a few days.

Hydrochloric acid is provided to the reaction by first treating the plate with a solution of magnesium chloride, drying it, and then treating it with TCQ solution. The plate is then heated at 110 °C for 10 min. The magnesium chloride reacts with the available water in the plate to form magnesium hypochlorite and hydrochloric acid and the reaction then proceeds. The co-ordinated P=S group gives reddish-orange colors and the bivalent P—S—C or C—S—C groups give yellow colors. If the plate is treated with a 2% potassium or sodium hydroxide solution and heated above 100 °C many compounds will hydrolyze to form phenols, these phenols can then couple with the TCQ to form the indophenol dye. This reagent has been quite useful for detecting the aromatic carbamate pesticides.

Another type of reaction that has been observed for TCQ is its formation of characteristically colored complexes with many compounds. These reactions may be addition type or charge-transfer type.[69] TCQ is therefore an example of a reagent which can be used for detecting different types of compounds, and which gives a different color reaction with each type. This would not be possible without knowing something about the nature of the chemical reactions themselves.

7.9 DERIVATIZATION

Derivatization can be a useful technique for thin layer chromatography as well as for the other chromatographic techniques. If thin layer chromatography is the only analytical technique available for determining low molecular weight or volatile compounds, derivatives can be made that allow them to be analyzed by this technique. Further, if a compound does not carry a substituent group on the molecule which can be reacted to produce fluorescence or a color, such a group can be added by derivatization. The reactions to be described here were originally used for making solid derivatives for melting point determinations. However these reactions are quite suitable for the uses proposed above. Derivatization reactions should generally be carried out after the environmental sample has been cleaned up (this extra treatment of the sample is also likely to reduce the levels of any interfering compounds). The stoichiometry of the reactions would need to be determined for the microgram range.

7.9.1 Alcohols and phenols

They can be derivatized by reacting with 3,5-dinitrobenzoyl chloride.[70]

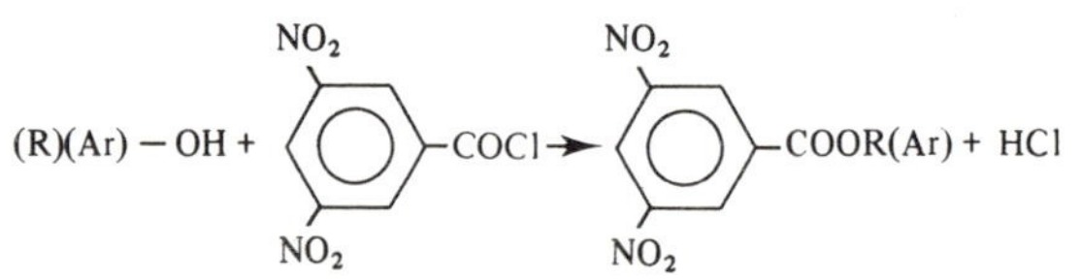

The derivatives can be further reacted with α-naphthylamine which may result in colored compounds.[71] They can either be chromatographed first, and then reacted with the amine or reacted first and then chromatographed. Similar reactions are obtained with *p*-nitrobenzoyl chloride,[72] *p*-nitrobenzyl bromide[73] for phenols and *p*-nitrophenyl isocyanate.[74]

Aliphatic alcohols react with 3-nitrophthalic anhydride[75] in the following manner:

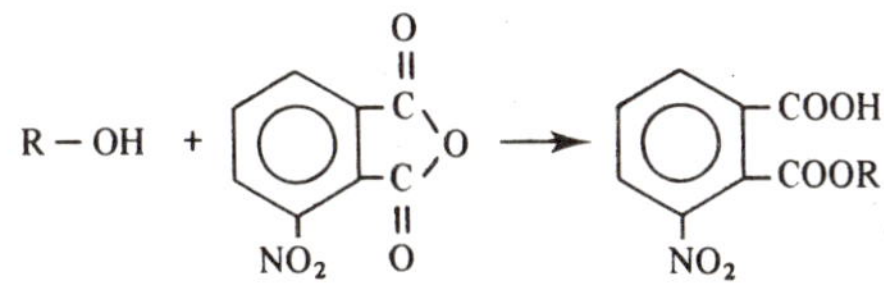

7.9.2 Aldehydes and ketones

The lower aldehydes and ketones will react with 2,4-dinitrophenylhydrazine to give yellow or red compounds.[76] They will also react with *p*-nitrophenylhydrazine.[77]

$$RR'CO + H_2N{-}HN{-}C_6H_4{-}NO_2 \rightarrow RR'C{=}N{-}HN{-}C_6H_4{-}NO_2 + HOH$$

Thiosemicarbazide reacts with aldehydes and ketones[78] to form thiosemicarbazones, which can form complexes with the metals silver, copper and mercury.

$$RR'C{=}O + H_2N{-}NH{-}CS{-}NH2 \rightarrow RR'C{=}N{-}NH{-}CS{-}NH_2 + HOH$$

$$RR'{-}C{=}N{-}NH{-}CS{-}NH_2 + M^+ \rightarrow RR'{-}C{=}N{-}N{=}C{-}\overset{M}{S}{-}NH_2 \quad \downarrow$$

This treatment with the metal salt after chromatography proves useful as a detection and identification procedure.

7.9.3 Carboxylic acids

Aniline will react directly with carboxylic acids to form anilides which can be chromatographed directly.[79]

$$R{-}COOH + C_6H_5NH_2 \rightarrow C_6H_5{-}NH{-}OC{-}R + HOH$$

The anilides can be broken down to aniline and carboxylic acid after chromatography by heating with an ammonia-contributing compound. The aniline may then be coupled to form a number of dyes.[80] Toluidine can be reacted in a similar manner. Nitrobenzyl bromide will react with carboxylic acids[81] as well as with phenols.

7.9.4 Amines

Primary amines also react with 2,4-dinitrochlorobenzene as do the secondary amines. Both aliphatic and aromatic amines will give positive reactions.[82]

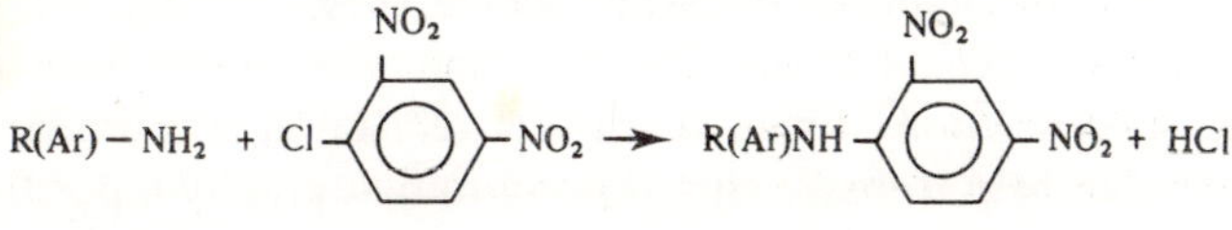

The aromatic primary amines will react with *p*-nitrobenzaldehyde[83] and give yellow colored derivatives.

$$ArNH_2 + OHC{-}C_6H_4{-}NO_2 \rightarrow ArN{=}HC{-}C_6H_4{-}NO_2 + HOH$$

Phthalic anhydride will also react with primary amines.[22]

The trinitro-substituted phenyl compounds are capable of some unique reactions, but they require special handling. They should be handled in a fume cupboard with a safety screen in place. The dilute solutions applied to thin layer plates during chromogenic reactions should not present an explosion hazard. However, when derivatives are made before chromatography, the solutions involved are more concentrated, and could present a hazard. Any raising of the temperature should be done in gradual stages instead of in a sudden surge.

Picric acid (2,4,6-trinitrophenol) can form picrates with aromatic hydrocarbons amines, phenols, alkaloids, heterocyclic compounds and aromatic ethers. The compounds are probably addition complexes.[84] Picryl chloride (2,4,6-trinitrochlorobenzene) reacts with both primary and secondary amines.[85]

7.9.5 General addition reagents

These compounds are brightly colored, and therefore self-indicating.

(i) *Styphnic acid* (2,4,6-trinitroresorcinol) will form addition compounds with hydrocarbons and alkaloids.[86]

(ii) sym-*Trinitrobenzene* (1,3,5-trinitrobenzene) forms addition products with aromatic hydrocarbons, primary aromatic amines, tertiary amines, heterocyclic compounds, phenols and aromatic ethers.[87] These products invariably have bright colors.

7.9.6 Other derivatization techniques

Any chemical reaction which will convert material separated by thin layer chromatography into a form in which it is more readily detected, may be used. A comprehensive handbook of such reactions has recently been published by Blau and King.[88]

7.10 IDENTIFICATION

A number of chromogenic reagents have been discussed in the foregoing section and their mechanisms described where they were known. Several of the reactions have been known for years and some have been applied to colorimetry. The reactions which selectively rely upon substituent groups in the molecules can help to make thin layer chromatography a positive identification tool. When unknown environmental samples are analyzed it is certainly helpful to be able to characterize some of the groups in the molecule.

If, however, an analyst does not wish to be concerned with the chemistry of reactions, information on appropriate and satisfactory reagents can be gleaned from several good sources. Books by Blau and King,[88] Hais and Macek,[89] Macek,[90] Stahl[91] and Zweig and Sherma[92] list many reagents along with the compounds they are used to detect. This does not guarantee that the investigator will get good results at the first attempt. Often, these procedures require modification because of differences in adsorbent binders or impurities in the solvent system.

7.11 DETECTION SENSITIVITY

The author has eliminated sensitivity tables from this text because they are very subjective, and sometimes sensitivity instead of reliability becomes the object of research. The lower limit of sensitivity of any test should be determined experimentally by the analyst applying the procedure.

One way to determine sensitivity is to chromatograph 0.05 to 2 μg concentrations of the standard and react them with the reagent. This, of course, only gives the sensitivity of the reagents to the standard. When an actual environmental sample containing the lowest level of a compound that could possibly give a visible reaction is analyzed, it is often found that the compound can no longer be seen. A law of diffusion seems to apply for environmental samples, though no absolute recovery levels can be postulated.

Isolation procedures usually have to be more rigorous where larger samples are spotted. Oil-type interferences can make a paper or thin layer chromatogram worthless.

7.12 BIOAUTOGRAPHY

Bioautography is quite useful as a sensitive and selective technique for paper and thin layer chromatography. Either live organisms or enzymes are used as detecting agents for the chromatographed compounds. Live organisms can be used as detectors for antibiotics, fungicides and other types of biochemical inhibitors. The chromatogram is treated with a solution containing the organism and, after a period of incubation, zones where their growth is inhibited become apparent. These zones are usually clear areas against the opaque or cloudy background produced by the growing organism.

The direct application of live organisms to the chromatogram has not always proved satisfactory, so many investigators use the transfer method. This involves casting a slab of starch, agar or some other type of gel which has already been seeded with a particular organism. The slab is then placed over the chromatogram, in contact with the chemicals on its surface, and incubated long enough to show the zones of inhibition.

Billow and Speaker[93] used precoated chromatographic sheets of silica gel, alumina, polyamide or cellulose, sandwiched between a thin layer of gelled nutrient and an overlying layer of nutrient agar seeded with the test organism. They used *E. coli* and *S. aureus* as the test organisms and, before chromatography, the plate was irradiated with UV radiation for 1 h to sterilize the surface and minimize contamination by organisms other than those being tested.

The sheets were scored into channels 2 cm wide, and were cut apart after development to form ten 2 cm wide sheets. A base layer of 100 ml of nutrient agar was poured into a sterile Pyrex baking dish (22 × 33 × 4 cm). When the agar had almost congealed the chromatographic strips were placed on the agar surface with the adsorbent layer facing up. The agar was then allowed to congeal completely. The strips were then overlaid with 200 ml of seeded agar. After the seeded agar had congealed, the dish was covered with kraft paper and incubated at 37 °C for 24 h. The antibiotic-containing areas were visible as clear zones on agar beds otherwise covered with bacterial growth.

Begue and Kline[94] enhanced the visibility of inhibition zones by reacting the bacterial growth with tetrazolium salts after incubation with the chromatogram. The seeded agar slab was incubated with the chromatogram and then sprayed with an aqueous tetrazolium salt solution. Various results were obtained depending upon the organism in the agar slab. The authors reported the best results with *p*-iodonitrotetrazolium violet, where sharp zones of bacterial growth inhibition were observed against a colored background.

Enzymes have also been used for detecting subnanogram levels of pesticides. In particular, the cholinesterase-inhibiting pesticides have been detected with great success. The enzyme can be obtained from human serum, horse serum, beef liver, bee heads or from commercially isolated acetylcholin esterase.

Getz and Friedman[95] were able to react organophosphate pesticides directly with the enzyme and substrate on a paper chromatogram to reveal the inhibiting pesticides as blue spots against a yellow background. The paper chromatograms were sprayed with a solution of human serum, base and bromthymol blue, incubated for 30 min, and then sprayed with a substrate solution of acetylcholine. The enzyme in the background should hydrolyze the acetylcholine, liberating acetic acid which turns the bromthymol blue yellow. Sites containing the migrated pesticide inhibit the hydrolytic enzyme and are therefore visible as blue areas.

When interferences were known to be present on the chromatogram, a separate sheet of chromatographic paper was sprayed with the serum indicator solution and held in close contact with the chromatogram for 30 min. The sheet was then lifted off and sprayed with the substrate to reveal the pesticides. The transfer could be carried out several times.

When this proccedure was applied to thin layer chromatograms silica gel proved too acid and tended to turn the plate yellow; alumina proved too basic, so that the background remained blue despite the action of the enzyme. However, reactions with cellulose thin layer plates were found to be satisfactory.

Mendoza[96] has written two comprehensive review articles which cover quite well the application of enzyme inhibition to the detection of pesticides.

There is, however, still a great deal of research to be done on the application of microorganisms and enzymes as chromogenic reagents in chromatography. Although they can be made highly selective and sensitive, it is still desirable to develop simpler and more direct techniques.

7.13 QUANTITATIVE THIN LAYER CHROMATOGRAPHY

Thin layer chromatography is no longer simply a qualitative technique. It can be as quantitative as any other chromatographic method because *in situ* determination of spot intensity or spot size is now possible. The plate is optically scanned using either reflected or transmitted light. A standard curve is plotted from the reference standards and the amount of an unknown present is determined from it. Concentration is usually plotted against the area of the curve produced on a recorder. The technique is an example of *in situ* spectrophotometry, for which several excellent scanners are now commercially available, with instrumentational capabilities ranging from the very simple to the highly complex. Indeed, using techniques such as these it is possible to adapt most of the methods described in this book for quantitative assay, so that environmental analysts now have three quantitative chromatographic methods at their disposal, viz., gas/liquid chromatography, liquid/liquid chromatography and thin layer chromatography, all of which can be highly effective tools for monitoring chemicals in the environment.

REFERENCES

1. R. J. Block, E. L. Durrum and G. Zweig, *A Manual of Paper Chromatography and Paper Electrophoresis*, Academic Press, New York, 1955.
2. I. M. Hais and K. Macek, *Handbuch der Papierchomatographie*, Fischer Verlag, Jena, DDR, 1958.
3. I. Smith, *Chromatographic and Electrophoretic Techniques*, Vol. I, Interscience, New York, 1960.
4. M. E. Getz, *J. Assoc. Off. Anal. Chem.* **54**, 982 (1971).
5. S. C. Chang, private communication, Chemosterilant Laboratory, AEQI, ARS, US Dept of Agriculture, Beltsville, Md. USA (1977).
6. F. Geiss and H. Schlitt, *Naturwissenschaften* **50**, 1350 (1963); F. Geiss, H. Schlitt and A. Lose, *Z. Anal. Chem.* **213**, 321, 331 (1965).
7. R. A. De Zeeuw, *J. Chromatogr.* **32**, 43 (1968); R. A. De Zeeuw, *Anal. Chem.* **40**, 915 (1968).
8. J. A. Perry, K. W. Haag and L. J. Glunz, *J. Chromatogr. Sci.* **11**, 447 (1973).
9. L. C. Mitchell, *J. Assoc. Off. Agric. Chem.* **40**, 999 (1957).
10. T. Beesley, *Kontes Technical Bulletin No. 201*, Kontes, Vineland, NJ, USA (1976).
11. W. Bradley and J. D. Thompson, *Chimia* **15**, 147 (1961).

12. J. J. Randall, C. E. Lewis and P. M. Slagan, *J. Org. Chem.* **27**, 4098 (1962).
13. A. C. Bratton and E. I. Marshall, *J. Biol. Chem.* **128**, 537 (1939).
14. E. Sawicki, T. W. Stanley and T. R. Hauser, *Chem. Anal.* **48**, 30 (1959).
15. R. R. Philips, *Org. React.* **10**, 143 (1959).
16. P. W. Vittum and A. Weissberger, *J. Photogr. Sci.* **2**, 81 (1954).
17. J. T. Stewart, T. D. Shaw and P. B. Ray, *Anal. Chem.* **41**, 360 (1969).
18. H. O. Friestad, D. E. Ott and F. A. Gunther, *Anal. Chem.* **41**, 1750 (1969).
19. E. F. Mohler Jr and L. N. Jacob, *Anal. Chem.* **29**, 1369 (1957).
20. J. Epstein, R. W. Rosentjal and R. J. Ess, *Anal. Chem.* **27**, 1435 (1955).
21. F. Feigl, *Spot Tests in Organic Analysis*, Elsevier, Amsterdam, 1966.
22. R. H. F. Manske, *J. Am. Chem. Soc.* **51**, 1202 (1929).
23. F. Feigl, *Angew. Chem.* **73**, 656 (1961).
24. F. Feigl and C. Stark, *Anal. Chem.* **27**, 1838 (1955).
25. F. Feigl and J. R. Amaral, *Anal. Chem.* **30**, 1148 (1958).
26. M. E. Getz, *J. Assoc. Off. Agric. Chem.* **47**, 1103 (1964).
27. F. Feigl, V. Anger and R. Zappert, *Mikrochemie* **15**, 12 (1934).
28. S. Ohkuma, *J. Pharm. Soc. Jpn* **75**, 1124 (1955).
29. F. Feigl, *Anal. Chem.* **27**, 1315 (1955).
30. F. Feigl, *Spot Tests in Organic Analysis*, Elsevier, Amsterdam, 1966, p. 264–265.
31. V. Anger and S. Ofri, *Mikrochim. Acta* **27**, 627 (1964).
32. E. Koenig and H. Greiner, *Ber.* **64**, 1049 (1931).
33. F. Feigl and E. Libergott, *Anal. Chem.* **36**, 132 (1964).
34. F. Feigl, L. Ben-Dor and E. Jungreis, *Chem. Anal.* **52**, 113 (1963).
35. V. Anger and S. Ofri, *Z. Anal. Chem.* **203**, 425 (1964).
36. J. Tröger and O. Müller, *J. Prakt. Chem.* **78**, 371 (1908).
37. F. Feigl, V. Anger and G. Fischer, *Mikrochim. Acta* **1**, 127 (1937).
38. R. Wasicky and O. Frehden, *Mikrochim. Acta* **1**, 55 (1937).
39. F. Feigl and R. A. Rosell, *Z. Anal. Chem.* **159**, 335 (1958).
40. V. M. Adamovic, *J. Chromatogr.* **23**, 274 (1966).
41. A. L. Le Rosen, R. T. Moravek and J. K. Carlton, *Anal. Chem.* **24**, 1335 (1952).
42. F. Feigl and E. Jungress, *Anal. Chem.* **31**, 2099 (1959).
43. P. Ehrlich, *Z. Klin. Med.* **5**, 285 (1882).
44. P. Ehrlich and C. A. Herter, *Z. Physiol. Chem.* **41**, 329 (1904).
45. K. Kobert, *Z. Anal. Chem.* **46**, 711 (1907).
46. F. Buscarons, J. C. Marins and J. Claver, *Anal. Chim. Acta* **3**, 319 (1949).
47. F. Feigl, *Spot Tests in Organic Analysis*, Elsevier, Amsterdam, 1966, pp. 421–422.
48. F. Feigl, *Spot Tests in Organic Analysis*, Elsevier, Amsterdam, 1966, pp. 192–193.
49. F. Feigl, V. Anger and G. Fischer, *Mikrochim. Acta* **25**, 878 (1962).
50. L. Rosenthaler and G. Vegezzi, *Mitt. Geb. Lebensmittelunters. Hyg.* **45**, 178 (1954).
51. H. Wieland and G. Scheuing, *Ber.* **54**, 2527 (1921).
52. V. Anger and G. Fischer, *Mikrochim. Acta* **23**, 592 (1960).
53. F. Feigl, V. Anger and O. Frehden, *Mikrochemie*, **15**, 12 (1934).
54. F. Feigl, *Spot Tests in Organic Analysis*, Elsevier, Amsterdam, 1966, pp. 22–23.
55. F. Feigl, *Spot Tests in Organic Analysis*, Elsevier, Amsterdam, 1966, pp. 229–230.
56. C. Lieberman, *Ber.* **7**, 247 (1874).
57. F. Feigl, V. Anger and O. Frehden, *Mikrochemie* **15**, 181 (1934).
58. V. Anger, *Michrochim. Acta* **23**, 827 (1960).
59. F. Feigl and Cl. Costa Neto, *Anal. Chem.* **28**, 1311 (1956).
60. F. Feigl, *Spot Tests in Organic Analysis*, Elsevier, Amsterdam, 1966, p. 294.
61. F. Feigl, *Spot Tests in Organic Analysis*, Elsevier, Amsterdam, 1966, p. 295.
62. P. K. Bose, *Analyst* **56**, 504 (1931).
63. T. Pohloudek-Fabini, D. Göckeritz and M. Schüssler, *Mikrochim. Acta* **26**, 668 (1963).

64. F. Feigl, V. Gentil and D. Haguenauer-Castro, *Microchem J.* **4**, 445 (1960).
65. H. D. Gibbs, *J. Biol. Chem.* **72**, 649 (1927).
66. E. Leger, *Z. Anal. Chem.* **38**, 374 (1889).
67. M. T. Cuzzoni and T. Pietra Lissi, *Farmaco Ed. Sci.* **19**, 981 (1964).
68. M. Beroza, K. R. Hill and K. H. Norris, *Anal. Chem.* **40**, 1608 (1968).
69. M. E. Getz, unpublished work.
70. W. N. Lipscomb and R. H. Baker, *J. Am. Chem. Soc.* **64**, 179 (1942).
71. M. Philips and G. L. Keenan, *J. Am. Chem. Soc.* **53**, 1924 (1931).
72. D. W. Adamson and J. Kenner, *J. Chem. Soc.* **141**, 286 (1935).
73. J. A. Lyman and E. E. Reid, *J. Am. Chem. Soc.* **42**, 615 (1920).
74. R. L. Shriner and R. F. B. Cox, *J. Am. Chem. Soc.* **53**, 1601 (1931).
75. B. H. Nicolet and J. Sachs, *J. Am. Chem. Soc.* **47**, 2348 (1925).
76. H. A. Iddles, A. W. Low, B. D. Rosen and R. T. Hart, *Ind. Eng. Chem. Anal. Ed.* **11**, 102 (1939).
77. H. Behrens, *Chem. Ztg.* **27**, 1105 (1903).
78. R. Fischer and A. Nour, *Mikrochem.* **15**, 1210 (1919).
79. P. W. Robertson, *J. Chem. Soc.* **115**, 1210 (1919).
80. F. Feigl, *Spot Tests in Organic Analysis*, Elsevier, Amsterdam, 1966, pp. 260–261.
81. E. E. Reid, *J. Am. Chem. Soc.* **39**, 124 (1917).
82. M. Willgerodt, *Ber.* **9**, 977 (1876).
83. *Organic Reagents for Organic Analysis*, Chemical Rubber Publishing Co., Brooklyn, New York, 1946, pp. 65–66.
84. *Organic Reagents for Organic Analysis*, Chemical Rubber Publishing Co., Brooklyn, New York, 1946, pp. 86–87.
85. J. C. Crocker and F. Matthews, *J. Chem. Soc.* **99**, 301 (1911).
86. C. S. Gibson, *J. Chem. Soc.* **93**, 2098 (1908).
87. H. Hibbert and J. J. Sudborough, *J. Chem. Soc.* **83**, 1334 (1903).
88. *Handbook of Derivatives for Chromatography*, Ed. by K. Blau and G. King, Heyden, London, 1977.
89. I. M. Hais and K. Macek, *Paper Chromatography*, Academic Press, New York, 1963.
90. *Pharmaceutical Applications of Thin-Layer and Paper Chromatography*, Ed. by K. Macek, Elsevier, New York, 1972.
91. *Thin-Layer Chromatography*, 2nd Edn, Ed. by E. Stahl, Springer Verlag, New York, 1969.
92. *Handbook of Chromatography*, Vol. II, Ed. by G. Zweig and J. Sherma, CRC Press, Cleveland, Ohio, 1972.
93. J. Billow and T. J. Speaker, *J. Chromatogr.* **67**, 191 (1972).
94. W. J. Begue and R. M. Kline, *J. Chromatogr.* **64**, 182 (1972).
95. M. E. Getz and S. J. Friedman, *J. Assoc. Off. Agric. Chem.* **46**, 707 (1963).
96. C. E. Mendoza, *Residue Rev.* **43**, 105 (1972); **50**, 43 (1974).

INDEX